Advanced Manufacturing Processes

Advanced Manufacturing Processes

Gustavo H. S. F. L. Carvalho
Rui Manuel Leal
Ivan Galvão

MDPI

Basel • Beijing • Wuhan • Barcelona • Belgrade • Novi Sad • Cluj • Manchester

Gustavo H. S. F. L. Carvalho
Department of Industrial Engineering
University of Florence
Florence
Italy

Rui Manuel Leal
CEMMPRE/University of Coimbra
Coimbra
Portugal

Ivan Galvão
UnIRE, Department of Mechanical Engineering
ISEL/Polytechnic University of Lisbon
Lisbon
Portugal

Editorial Office
MDPI AG
Grosspeteranlage 5
4052 Basel, Switzerland

This is a reprint of articles from the Special Issue published online in the open access journal *Applied Sciences* (ISSN 2076-3417) (available at: www.mdpi.com/journal/applsci/special_issues/Advanced_Manufacturing_Processes).

For citation purposes, cite each article independently as indicated on the article page online and using the guide below:

Lastname, A.A.; Lastname, B.B. Article Title. *Journal Name* **Year**, *Volume Number*, Page Range.

ISBN 978-3-7258-2076-4 (Hbk)
ISBN 978-3-7258-2075-7 (PDF)
https://doi.org/10.3390/books978-3-7258-2075-7

Cover image courtesy of Gustavo H.S.F.L. Carvalho; Rui M. Leal; Ivan Galvão

Contents

About the Editors

Gustavo H. S. F. L. Carvalho

Gustavo Senna Carvalho is an assistant professor at the Department of Industrial Engineering of the University of Florence (Italy) and collaborator researcher at CEMMPRE—University of Coimbra (Portugal). He has a PhD degree in Mechanical Engineering (University of Coimbra), a Master's degree in Metallurgical and Materials Engineering (Federal University of Rio de Janeiro), and a Bachelor's degree in Mechanical Engineering (CEFET/RJ). He has been working on several topics within manufacturing technologies, such as welding processes, welding metallurgy, additive manufacturing, and material processing. The main research topics are solid state welding processes (mainly explosion welding and friction stir welding), cladding, and wire arc additive manufacturing based on the metallurgical and microstructural analysis and their relation to the mechanical properties, the welding quality, and the component's integrity. He has published many articles in international scientific journals, is a reviewer for several international journals, and periodically participates in national and international conferences and events.

Rui Manuel Leal

Rui Manuel Leal has a PhD in Mechanical Engineering (Production Technology field), an MSc in Mechanical Engineering (Technology and Materials field), and a BSc in Mechanical Engineering (Production field) from the Faculty of Sciences and Technology at the University of Coimbra (Portugal). He is currently an assistant professor at ESAD.CR—Polytechnic University of Leiria, where he teaches curricular units of technology, materials, and technical drawing. He is also an integrated researcher at the CEMMPRE, University of Coimbra, and a collaborative researcher at the Research Laboratory in Design and Arts (LIDA). His main research areas are focused on production technologies, having collaborated on several research projects. As a result of his research, he has participated in national and international events, such as conferences and congresses, and has published several works in scientific journals.

Ivan Galvão

Ivan Galvão is an assistant professor at the Department of Mechanical Engineering of ISEL—Polytechnic University of Lisbon (Portugal) and researcher at UnIRE— Polytechnic University of Lisbon (Portugal) and CEMMPRE—University of Coimbra (Portugal). He has a PhD in Mechanical Engineering (Production Technology field) from the University of Coimbra (Portugal). His research field comprises the joining and processing of materials, specifically, welding technology and metallurgy, the mechanical behavior of welded joints, the solid-state welding of similar and dissimilar materials, and the solid-state processing of materials. He has published many scientific papers in high impact journals and conference proceedings, and he has participated in several international conferences. Other activities developed by him include his involvement in research projects, his collaboration with international journals and scientific and engineering associations, and his supervision of research works and theses.

applied sciences

MDPI

Editorial

Advanced Manufacturing Processes

Gustavo H. S. F. L. Carvalho [1,2,*], Rui M. Leal [2,3] and Ivan Galvão [2,4]

1 Department of Industrial Engineering, University of Florence, 50121 Florence, Italy
2 CEMMPRE, Department of Mechanical Engineering, University of Coimbra, 3030-788 Coimbra, Portugal; rui.leal@ipleiria.pt (R.M.L.); ivan.galvao@isel.pt (I.G.)
3 LIDA-ESAD.CR, Polytechnic University of Leiria, 2500-321 Caldas da Rainha, Portugal
4 UnIRE, ISEL, Polytechnic University of Lisbon, 1959-007 Lisbon, Portugal
* Correspondence: gustavo.carvalho@unifi.it

1. Introduction

The advancement of engineering and the emergence of increasingly individualized and specific applications have heightened the need to develop advanced manufacturing processes capable of delivering all of the required characteristics. These processes, which have very particular features or have been developed more recently, deserve special attention in current manufacturing engineering research in order for their challenges to be overcome. This progress will spread the use of these processes, resulting in better-quality components with lower manufacturing costs and lower environmental impact [1].

Processes such as welding and joining are constantly improving and they remain a crucial technology for modern engineering [2,3]. In turn, subtractive manufacturing is one of the most important steps in the manufacturing of a component and represents an essential part of manufacturing, even with the development of additive processes [4,5]. Material forming and assembly are also topics of interest that still have their areas of relevance in the current manufacturing industry [6,7]. In fact, component manufacturing is so varied and diverse that each manufacturing process has its own space and application.

The aim of this Special Issue is to illustrate the state of the art of advanced manufacturing processes, giving a unique relevance to recent developments achieved in this field, which may provide innovative engineering solutions to the industry.

Citation: Carvalho, G.H.S.F.L.; Leal, R.M.; Galvão, I. Advanced Manufacturing Processes. *Appl. Sci.* **2024**, *14*, 7553. https://doi.org/10.3390/app14177553

Received: 16 July 2024
Accepted: 13 August 2024
Published: 27 August 2024

2. Overview of Published Articles

The present Special Issue encompasses 11 original research papers, each addressing different areas of manufacturing technology. Specifically, four-papers address material welding and joining, five papers are focused on subtractive manufacturing (machining/cutting processes), one paper addresses material forming, and one paper focuses on dimensional control in assembly. Different research topics and different methodological approaches have been studied/followed among the works published in this Special Issue, highlighting its multidisciplinary nature.

The works addressing welding and joining processes encompass studies on solid-state welding techniques, like friction welding, friction stir welding and magnetic pulse welding, and fusion welding techniques, specifically laser welding. It is important to stress that all of these works address the joining of dissimilar materials with very different chemical, physical, and mechanical properties. Regarding the study on friction welding, Winiczenko et al. (Contribution 1) used this technique to produce dissimilar welds of tungsten-heavy alloy and 5XXX aluminum alloy. The research approach followed by these authors was based on experimental and finite-element results. A good correlation between the experimental and the numerical results was found in terms of the weld geometric characteristics and temperature. The maximum weld temperature was found to be 581 °C, which was slightly higher than the temperature measured during the experimentation. The flash width, flash height, and axial weld shortening were predicted with small deviation values, which ranged between 2.34% and 4.45%.

Also using a friction-based welding technique, Torabi et al. (Contribution 2) addressed the dissimilar friction stir welding of aluminum and magnesium. These authors explored the influence of tool rotation speed on the formation of eutectic structures during the friction stir welding of aluminum to magnesium in order to avoid liquation and solidification problems. Through many characterization techniques, they found that changing the rotation speed directly affected the eutectic formation, whereas the welding speed did not present any influence. A lower rotation speed during the process resulted in a thin, continuous intermetallic layer, whereas a higher speed led to the formation of a massive Mg-$Al_{12}Mg_{17}$ eutectic microstructure. The formation of a eutectic mixture, as indicative of liquation, may affect the material flow during the process due to the decreased friction coefficient between the tool and material.

Similar to the previous two works, a solid-state welding technique was also studied by Shim and Kim (Contribution 3). However, in order to produce Al/steel joints, these authors used an impact-based technique, i.e., magnetic pulse welding. In fact, according to these authors, the use of evolutionary algorithms to reach optimal process parameters in magnetic pulse welding has been poorly explored in the literature. This way, their work was dedicated to maximizing the quality of Al/steel magnetic-pulsed (MP) welds by using evolutionary algorithms, i.e., an Imperialist Competitive Algorithm (ICA) and a Genetic Algorithm (GA). Prediction models were developed; the peak current, gap between workpieces, and frequency were selected as the input parameters; and the maximum load and weld length were selected as the output parameters. Their results showed a good agreement between the predicted values and experimental results for the maximum load and weld length, mainly when the ICA-based model was used.

Unlike the welding works described above, Wang et al. (Contribution 4) worked on fusion welding—in particular, on laser direct joining. Specifically, they studied the influence of temperature distribution on the interfacial bonding process between carbon-fiber-reinforced thermoplastic (CFRTP) composites and 6061 aluminum alloy during laser direct joining. To better characterize the interfacial bonding phenomena, these authors developed a novel online observation device that allowed them to observe the polymer melting, flowing, and bonding with the metal throughout the process. The temperature distribution/gradient in the weld zone was characterized using numerical simulation. Ultimately, the experimental and simulated results allowed the authors to correlate the temperature distribution with the melting and flow of the polymer. The melting of CFRTP composites and the formation of bubbles in the polymer matrix, which are associated with thermal degradation of the polymer and hinder its melting, could be determined by the temperature distribution in the bonding zone. This study also led the authors to conclude that the flow of the polymer during welding is responsible for thermal gradients. Moreover, the flow of the polymer was found to have a close relationship with the greater or lesser formation of bonding defects.

A significant number of works addressing subtractive manufacturing are published in the present Special Issue. These works encompass different processes, such as turning, milling, sawing, and grinding, with varying working scales (micro vs. macro). For example, the objective of the research developed by González-Sierra et al. (Contribution 5) was to evaluate the wear of ceramic and coated carbide inserts in the finishing turning of age-strengthened gray cast iron. Several experimental characterization techniques were used to analyze the wear of the inserts and the microstructure of the workpiece materials (i.e., 5- and 12-day age-strengthened gray cast iron). The authors observed a slight increase in the tensile strength of each workpiece material by increasing the degree of age strengthening from 5 to 12 days. They also report a 50% reduction in flank wear for inserts that machined 5-day age-strengthened cast iron when compared with inserts that machined the cast iron with a greater degree of age strengthening. The main wear mechanisms in the inserts were found to vary according to the insert type—while abrasion and adhesion were the main wear mechanisms of ceramic inserts, adhesion and oxidation were the main wear mechanisms for coated carbide tools.

Jia et al. (Contribution 6) contribute to this Special Issue with a paper on diamond wire sawing, which is the most widely used method of cutting Al_2O_3 ceramic due to its high accuracy and the minimal surface damage it causes. Due to the need to improve the material removal rate, park discharges were generated around the diamond wire based on the electrochemical discharge machining (ECDM) process. An oil-film-assisted electrochemical discharge machining process was applied to solve the difficulty of generating spark discharges. In this work, the authors found that the combination of oil-film-assisted electrochemical discharge machining and diamond wire sawing improved the material removal rate of Al_2O_3 ceramic, improved the surface quality of the machined parts, and reduced the wear on the diamond wire. The results of their study bring valuable information for the application of diamond wire sawing combined with oil-film-assisted ECDM.

An abrasive material removal technique was addressed by Wang et al. (Contribution 7), who studied the surface quality of large-shaft multi-pass grinding. Two of the most important surface quality indices (surface roughness and glossiness) are affected by the process parameters and the surface quality of the previous grinding pass, which leads to difficulty in modeling. Moreover, an uneven distribution of the actual grinding depth usually leads to inconsistent surface quality throughout the shaft, which results in the need for multiple spark-out grinding passes to ensure consistency. Thus, these authors developed surface quality evolution models for assessing surface roughness and glossiness based on an Elman neural network, which builds regressions between process parameters, surface quality indices of the previous grinding pass, and surface quality indices of the current grinding pass. These authors also propose a consistency control method for the surface quality that involves adjusting the actual grinding depth within the dimensional accuracy tolerance range at the rough grinding stage. Their study shows that this proposed consistency control method can guarantee a consistent surface quality, reduce the grinding passes, and increase the grinding efficiency.

At a lower scale than in the previous works on subtractive manufacturing, the paper published by Wang et al. (Contribution 8) addresses micro-milling, another particular process within engineering that presents many applications such as terahertz slow-wave structures, microfluidic chips, and micro-molds. These authors indicate that even with this process consuming more energy than traditional milling, research on micro-milling is usually focused on parameter optimization, with limited attention given to energy efficiency. Citing the importance of developing or improving such processes to be more sustainable, the authors propose a micro-milling parameter-based power consumption model for use with evolutionary algorithms to optimize micro-milling parameters. This model allowed them to achieve comprehensive enhancements in both machinability and sustainability, improving surface quality, dimensional accuracy, the material removal rate, and specific energy consumption during the micro-milling process for thin-walled micro-structures.

Wang et al. (Contribution 9) also worked on machining, but with a different focus, specifically the use of double-sided collaborative machining. In fact, these authors investigated the tool path in double-sided collaborative machining of a propeller blade. The concomitant vibration and deformation produced by propeller blades in single-sided machining significantly affect the surface machining precision. These authors indicate that double-sided symmetrical machining can improve system rigidity through mutual shoring on both sides, but it cannot be applied to blade machining due to the complexity of a blade's shape. Thus, they investigated a tool-path planning algorithm, showing that the algorithm can achieve smooth fitting and correspondence of bilateral cutter position points through double-curve interpolation and position data alignment. The authors also verified the feasibility and superiority of double-sided collaborative machining through machining experiments.

The multidisciplinary nature of the present Special Issue also allowed the inclusion of the work conducted by Zhang et al. (Contribution 10). These authors investigated the variations in surface shape during gas jet forming for optical aspherical mirror blanks. The inference was verified using numerical simulation and compared to the experimental results. The authors found that the average prediction deviation for the diameter was 1.07 mm, while the average prediction deviation for the principal curvature was 0.03665 mm^{-1}, which is challenging to correct in simulation. Therefore, they developed a dimensionless prediction model for forecasting the surface curvature and surface diameter of mirror blanks by considering the jet parameters using experimental data. The model's average prediction error for the surface diameter of the formed surface was found to be 0.3192 mm, and the average prediction error of the principal curvature for the formed surface was 0.00269 mm^{-1}.

The assembly stage of component manufacturing is also addressed in detail in this Special Issue. Zhang et al. (Contribution 11), taking the core machine of a micro gas turbine as the research object, conducted a study with the aim of constructing a simulation model for assembly-deviation prediction based on the design tolerance and actual measurement data. The authors analyzed the assemblability of the design model and the main factors promoting assembly deviation. The current assembly position was found to be non-optimized, which led the authors to propose an optimal solution, resolving the interference issue resulting from the design error. The authors also report an assembly deviation analysis showing that when using an assembly-deviation prediction and analysis model that is constructed based on actual measurement data, one can achieve results that exhibit high consistency with the actual outcome of the assembly.

3. Conclusions

This set of articles positively demonstrates the exact objective of this Special Issue: to address arguments regarding advanced and unconventional manufacturing processes in order to take a step forward in the development and improvement of knowledge surrounding these processes.

In a world that is increasingly valuing and urgently needs more sustainable and efficient processes, manufacturing processes require constant improvement and development—and components must be manufactured with better quality. To keep pace with this evolution, it is important to study not only the main manufacturing processes, but also processes that are typically less discussed. This Special Issue presents research addressing topics such as welding, material removal processes, numerical simulation, tool wear, and sustainability, among others. These are all extremely important processes that are not commonly addressed in the existing literature, despite the fact that the knowledge surrounding these topics requires improvement in both its amount and its quality—thus, this corroborates the relevance of this Special Issue.

Finally, it is important to highlight the fact that the publications in this Special Issue come from various research groups spread widely across the world. The level of expertise among the researchers who have contributed to this Special Issue is outstanding, constituting the excellence of this compilation of scientific articles for the manufacturing sector.

As the Editors of the present Special Issue, we are pleased with the success of this project. We are profoundly grateful to all of the authors, the reviewers, and the Editorial Office that contributed to the making of this Special Issue.

Funding: This research is sponsored by national funds through FCT—Fundação para a Ciência e a Tecnologia—under the project UIDB/00285/2020 and LA/P/0112/2020.

Conflicts of Interest: The authors declare no conflicts of interest.

List of Contributions

1. Winiczenko, R.; Skibicki, A.; Skoczylas, P. The Experimental and FEM Studies of Friction Welding Process of Tungsten Heavy Alloy with Aluminium Alloy. *Appl. Sci.* **2024**, *14*, 2038. https://doi.org/10.3390/app14052038.

2. Torabi, K.; Beygi, R.; Eisaabadi Bozchaloei, G.; da Silva, L.F.M. The Effect of Tool Rotation Speed on the Formation of Eutectic Structure during Friction Stir Welding of Aluminum to Magnesium. *Appl. Sci.* **2023**, *13*, 7133. https://doi.org/10.3390/app13127133.
3. Shim, J.; Kim, I. Evolutionary Algorithm to Optimize Process Parameters of Al/Steel Magnetic Pulse Welding. *Appl. Sci.* **2023**, *13*, 12881. https://doi.org/10.3390/app132312881.
4. Wang, Q.; Fu, R.; Wang, F.; Luo, C.; Li, J.; Jia, Z. Effect of Temperature Distribution on Interfacial Bonding Process between CFRTP Composite and Aluminum Alloy during Laser Direct Joining. *Appl. Sci.* **2023**, *13*, 11973. https://doi.org/10.3390/app132111973.
5. González-Sierra, N.E.; Flores Méndez, J.; Meraz-Melo, M.A.; Piñón Reyes, A.C.; Munoz-Hernandez, G.A.; Morales-Sánchez, A.; Moreno Moreno, M.; Minquiz, G.M. An Evaluation of the Tool Wear of Ceramic and Coated Carbide Inserts in Finishing Turning under the Influence of Age-Strengthening Gray Cast Iron. *Appl. Sci.* **2023**, *13*, 10248. https://doi.org/10.3390/app131810248.
6. Jia, Z.; Zhang, K.; Wang, J. Surface Quality of Al_2O_3 Ceramic and Tool Wear in Diamond Wire Sawing Combined with Oil Film-Assisted Electrochemical Discharge Machining. *Appl. Sci.* **2023**, *13*. https://doi.org/10.3390/app13159030.
7. Wang, L.; Fu, S.; Wang, D.; Li, X. Surface Quality Evolution Model and Consistency Control Method of Large Shaft Multi-Pass Grinding. *Appl. Sci.* **2023**, *13*, 1502. https://doi.org/10.3390/app13031502.
8. Wang, P.; Bai, Q.; Cheng, K.; Zhao, L.; Zhang, Y. Multi-Objective Optimization of Micro-Milling Parameters—The Trade-Offs between Machining Quality, Efficiency, and Sustainability in the Fabrication of Thin-Walled Microstructures. *Appl. Sci.* **2023**, *13*, 9392. https://doi.org/10.3390/app13169392.
9. Wang, R.; Ge, Y.; Guo, X. Investigation on Tool Path Planning Algorithm of Propeller Blade Double-Sided Collaborative Machining. *Appl. Sci.* **2023**, *13*, 6529. https://doi.org/10.3390/app13116529.
10. Zhang, X.; Wang, M.; Fu, W. The Variation of Surface Shape in the Gas Jet Forming. *Appl. Sci.* **2023**, *13*, 504. https://doi.org/10.3390/app13010504.
11. Zhang, N.; Wu, Y.; Yang, R.; Xie, G. Digital Simulation and Analysis of Assembly-Deviation Prediction Based on Measurement Data. *Appl. Sci.* **2023**, *13*, 12193. https://doi.org/10.3390/app132212193.

References

1. Lambiase, F.; Scipioni, S.I.; Lee, C.-J.; Ko, D.-C.; Liu, F. A State-of-the-Art Review on Advanced Joining Processes for Metal-Composite and Metal-Polymer Hybrid Structures. *Materials* **2021**, *14*, 1890. [CrossRef] [PubMed]
2. Bharti, S.; Kumar, S.; Singh, I.; Kumar, D.; Bhurat, S.S.; Abdullah, M.R.; Rahimian Koloor, S.S. A Review of Recent Developments in Friction Stir Welding for Various Industrial Applications. *J. Mar. Sci. Eng.* **2024**, *12*, 71. [CrossRef]
3. Yang, Y.; Luo, Z.; Zhang, Y.; Su, J. Dissimilar Welding of Aluminium to Steel: A Review. *J. Manuf. Process.* **2024**, *110*, 376–397. [CrossRef]
4. Rahman, M.A.; Saleh, T.; Jahan, M.P.; McGarry, C.; Chaudhari, A.; Huang, R.; Tauhiduzzaman, M.; Ahmed, A.; Mahmud, A.A.; Bhuiyan, M.S.; et al. Review of Intelligence for Additive and Subtractive Manufacturing: Current Status and Future Prospects. *Micromachines* **2023**, *14*, 508. [CrossRef]
5. Lalegani Dezaki, M.; Serjouei, A.; Zolfagharian, A.; Fotouhi, M.; Moradi, M.; Ariffin, M.K.A.; Bodaghi, M. A Review on Additive/Subtractive Hybrid Manufacturing of Directed Energy Deposition (DED) Process. *Adv. Powder Mater.* **2022**, *1*, 100054. [CrossRef]
6. Miqueo, A.; Torralba, M.; Yagüe-Fabra, J.A. Lean Manual Assembly 4.0: A Systematic Review. *Appl. Sci.* **2020**, *10*, 8555. [CrossRef]
7. Trzepieciński, T. Recent Developments and Trends in Sheet Metal Forming. *Metals* **2020**, *10*, 779. [CrossRef]

applied sciences

MDPI

Article

The Experimental and FEM Studies of Friction Welding Process of Tungsten Heavy Alloy with Aluminium Alloy

Radosław Winiczenko [1,*], Andrzej Skibicki [2] and Paweł Skoczylas [3]

[1] Institute of Mechanical Engineering, Warsaw University of Life Sciences, Nowoursynowska 166, 02-787 Warsaw, Poland
[2] Faculty of Mechanical Engineering, Bydgoszcz University of Science and Technology, Kaliskiego 7, 85-789 Bydgoszcz, Poland; andrzej.skibicki@pbs.edu.pl
[3] Institute of Mechanics and Printing, Warsaw University of Technology, 02-524 Warsaw, Poland; pawel.skoczylas@pw.edu.pl
* Correspondence: radoslaw_winiczenko@sggw.edu.pl

Abstract: Experimental and finite element studies of the rotary friction welding (RFW) process of tungsten heavy alloy (THA) with aluminium alloy 5XXX series are presented. A 2.5D torsion simulation model including the circumferential effects was developed in this study. The temperature distributions, effective stress, flash dimensions and axial shortening were calculated on un-rotated friction welding aluminium parts. The peak temperatures were measured both in the axis and at the half-radius of the specimen. The maximum interface temperature of 581 °C for the friction weld was below the melting temperature of the aluminium alloy. The experimental and numerical results of the temperature and final weld geometries show good agreement between them. The results indicate very small deviations of 4.45%, 2.96%, and 2.34% on the flash width, flash height and axial shortening of friction welds.

Keywords: friction welding; tungsten heavy alloy; FEM

check for updates

Citation: Winiczenko, R.; Skibicki, A.; Skoczylas, P. The Experimental and FEM Studies of Friction Welding Process of Tungsten Heavy Alloy with Aluminium Alloy. *Appl. Sci.* **2024**, *14*, 2038. https://doi.org/10.3390/app14052038

Academic Editors: Gustavo H. S. F. L. Carvalho, Rui Manuel Leal and Ivan Galvão

Received: 29 December 2023
Revised: 2 February 2024
Accepted: 21 February 2024
Published: 29 February 2024

1. Introduction

Friction welding is a joining method that enables the relatively quick and cheap joining of various metals. The physical process of friction welding involves using mechanical energy and converting it into heat, as a result of friction on the contact surface. Friction welding is a solid-state welding method for different metals [1]. The material to be welded is heated to a temperature below its melting point. After stopping the spindle rotation, the joint is pressed and plasticized. As a result of the applied upsetting force, the material is removed to the outside where a flash weld is formed [2]. The friction welding process is a heat-activated process that causes the deformation of the welded material and a large temperature gradient in the heat-affected zone (HAZ). Moreover, large effective stresses are generated in the joint and the microstructural changes as a result of high temperatures at the interface [3].

The basic welding parameters in the direct-drive friction welding process are the rotational speed (RS), friction pressure (FP), upsetting pressure (UP), friction time (FT) and upsetting time (UT). The rotational speed is related to the material to be welded and the diameter of the specimens. The rotational speed also determines the heating rate at the interface. Increasing the rotational speed might lead to the softening of the material, a greater extent of recrystallisation, or even increased intermetallic formation. The maximum temperature in the joint during frictional heating depends not only on the pressure but also on the temperature gradient, which depends on the rotational speed in particular [2]. In rotary friction welding, heat is generated by friction, which results in the significant plasticization of the material. The friction pressure as well as subsequent upsetting during rotary friction welding cause changes in the microstructure of the joint and its surroundings, which can be compared with the changes caused by hot plastic working. A high temperature in some situations, especially when the process lasts longer, enables diffusion across the interface, resulting in a change in the chemical composition in the joint area. The parameters are selected depending on the type of materials to be welded. The friction pressure affects the heating rate of the HAZ. The greater the friction pressure, the higher the rate of plastic deformation in the welded materials and the higher the heating rate in the contact area. The friction time is set in close correlation with the friction pressure and rotational speed. This should ensure an optimal heating speed and the minimal shortening of the welded elements. The friction time is largely influenced by the thermal conductivity of the welded material. During the friction process, the greatest amount of heat is released around the circumference of the friction surface, where the metal achieves the greatest plasticity. The center of the friction surface is heated due to heat conduction. The larger the diameter and the lower the thermal conductivity of the metal, the lower the temperature at the center of the friction surface. The friction time should be selected so that the temperature is equalized over the entire friction surface. The upsetting pressure and time are parameters that mainly affect the quality of the joining materials. In rotary friction welding, the control of the peak and equilibrium torque levels are considered to be crucial for a sound weld. The axial pressure applied during friction welding will affect the rate of shortening (upset), which is usually constant during the equilibrium torque period. Moreover, the 'upset' can be linearly proportional to the axial pressure [3].

Many works have been dedicated to the thermomechanical processes that occur during friction welding. These thermal phenomena have most often been solved using a finite element method (FEM). Thermo-physical data and the material constants of plastic deformation are necessary for the modelling of the welding process.

The first extensive analysis of the conventional friction welding process was conducted in references [4,5]. The temperature distribution and uniform heat flow in the friction welding process of carbon steel bars with a constant friction coefficient and thermal properties were presented. The first numerical analysis of the friction welding process using the finite differences method was carried out by the authors of papers [6,7]. The authors analyzed the time-varying heat input, axial shortening and plastic deformation of welding materials. A two-dimensional finite-difference model for conducting simulations of the temperature fields of the friction welding process was also developed in references [8,9]. The results show a visible difference in the temperature distribution between the conventional and inertia friction welding (IFW) methods. The friction welding of steel bars with identical properties using the finite element method, along with the conditions of element contacts, was proposed in references [10,11]. Moreover, the temperature distribution and final shape of the welds were determined in references [12,13]. A thermo-mechanical finite element model of IFW of two bars as a viscoplastic material was proposed in reference [14]. The FEM model considered the plastic deformation of the joined workpiece and a heat flow analysis was conducted [15]. The effective stress, strain and temperature field for nickel-based alloy friction welded joints were determined. An FEM simulation of the IFW of materials with different physical properties that assumed mechanical and thermal dependencies and the elastic-viscoplastic behavior of the material was completed [16]. A new friction law for the modelling of the continuous drive friction welding (CDFW) of steel was proposed [17]. A FEM method for the friction welding of ceramic-metal [18] and refractory metals [19] was also used. The numerical model for a steady-state flow and the analysis of a coupled thermo-mechanical problem in friction welding was presented [20]. An investigation of the temperature distribution, axial shortening, flash formation and microstructure using inverse heat flow analysis can be found in reference [21].

Due to its high density, high strength, good toughness and other excellent properties, tungsten heavy alloy material has become the first choice for many kinds of kinetic energy armor-piercing bullets. Tungsten alloy penetrator bullets have a high penetration ability thanks to the high-density tungsten alloy, which makes the projectile fly at high speed and with high energy. Recently, tungsten heavy alloy projectile cores have been used as kinetic energy penetrators (KEPs) for sub-caliber bullets. These THA cores can be joined to a ballistic cap made of an aluminum alloy. This cap protects the projectile against ricochets when it hits the armor plate. These materials are often joined by threads, which require milling. This is problematic because tungsten is difficult to cut due to its relatively high hardness and strength. Therefore, the authors of references [22,23] have proposed the friction welding method for joining these materials. Friction welding is an effective solid-state joining method for various materials.

Compared to the conventional joining method, rotary friction welding is a simple, cheap and very fast process. Moreover, it makes it possible to join metals and alloys that are very difficult to join or are not joinable at all using other methods. This is why friction welding is a very attractive method and is studied in many laboratories across the world. However, there are some difficulties in joining dissimilar materials. These difficulties are caused by very different melting points, different thermal expansion coefficients, the forming of hard precipitates that are usually intermetallic phases, or problems caused by difficulties in the reaching of an appropriate temperature. To overcome these problems, many experimental works have been performed, especially for the joining of aluminum alloys with a stainless steel or tungsten composite.

The friction welding of THA with AA appears to be difficult due to their thermal properties and differing deformability characteristics [22,23]. When estimating the thermal field, a flash dimension and upset (axial shortening) in a friction-welded joint between these materials makes it much easier to control the welding parameters by reducing the number of necessary experiments. This work aims to develop a thermomechanical model using the finite element method to simulate the CDFW of a composite alloy with an aluminum alloy.

2. Methodology

2.1. Materials

Tungsten heavy alloy THA (Laboratory of Heavy Alloys, Institute of Mechanics and Printing, WUT, Warsaw, Poland) and aluminum alloy 5XXX series were used for the friction welding process. A commercial 5XXX series (AlMg3Mn)-type non-heat treatable wrought aluminum alloy and conventional W-Ni-Fe type tungsten heavy alloy (THA), with a typical 7:3 nickel to iron ratio, were selected for this study. The aluminum alloy used in this study was a cold-worked rod 20 mm in diameter. The tungsten heavy alloy was prepared by mixing the appropriate amount of powder, compacting it, and then using the liquid phase sintering (LPS) method. The method of producing the W–Ni–Fe alloy and aluminum rod is described in reference [23]. Because of the powder metallurgy method used for THA preparation, its microstructure consists of 30–40 μm tungsten grains embedded in a Ni(W, Fe) matrix. The chemical composition and properties of the joined materials are presented in Table 1.

Table 1. The properties of the materials used.

Chemical composition							
Material	Al	W	Fe	Ni	Mg	Mn	Si
THA	-	Bal.	2.25	5.25	-	-	-
AA	Bal.	-	0.194	0.002	2.95	0.265	0.1

Mechanical properties				
Material	Tensile Strength (MPa)	Yield strength (MPa)	Elongation (%)	Hardness (HB)
THA	960	680	27	285
AA	276	207	22	83

2.2. Experimental Procedure

A rotary drive friction welding machine (ZT4-13 type, ASPA, Wrocław, Poland) with the experimental setup of thermocouples shown in Figure 1 was used to produce a friction weld. Friction welding was achieved for the following parameters: friction force (FF = 12.5 kN), upsetting force (UF = 50 kN), friction time (FT) and upsetting time (UT) of 9.5 s and 6 s. The rotational speed was kept at a constant level of 1450 rpm. The simulations were performed for joints that showed the highest tensile strength of 232 MPa. The thermocouples (TP203K1b2001 type, 'Czaki Thermos-Product', Raszyn, Poland) with a diameter of approximately 1.2 mm were used to study the temperature distribution in the joint. Figure 1a shows a diagram of the arrangement of thermocouples in a non-rotating aluminum sample during welding. Figure 1b shows the measurement positions 1,2,3 for ½ R of the sample and 4,5,6 for the axis sample. These were located at a distance of 20, 25 and 30 mm from the weld interface. Temperature measurements were performed using the UT325 LCD digital recorder, Uni-Trend Technology, Dongguan City, Guangdong Province, China.

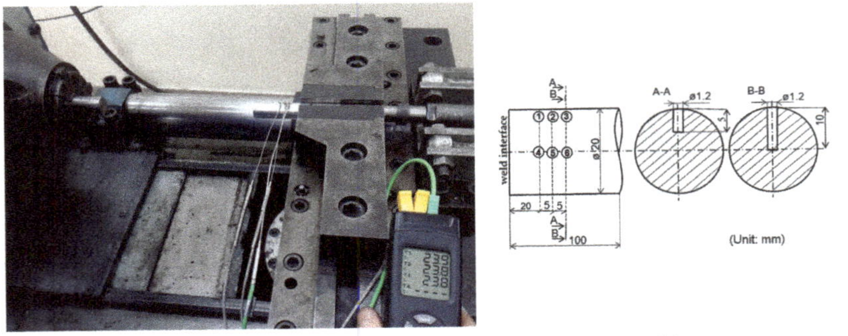

(a) (b)

Figure 1. Friction welding machine showing placed samples with thermocouple (**a**), and temperature measurement scheme (**b**) where: 1,2,3,4,5,6 are the thermocouples positions.

2.3. Numerical Procedure

The CDFW model was developed using the commercial finite element (FE) modeling software package, DEFORM 11.0. The software is especially capable of solving problems involving a large plastic deformation via remeshing [24]. The geometry and meshing of the FE model are shown in Figure 2.

A two-dimensional analysis of one-half of the solid element was employed because of the axisymmetric shape of the specimen (Figure 2a). The mesh contained 2125 elements and 2321 nodes (Figure 2b). The dimensions of the modelled parts are consistent with the experiment in terms of width. Three different mesh-sized windows were used. Coarse elements of 0.03 mm were used in the region containing the weld interface. Outside this region, as the distance increases, the element size also increases, and a mesh size of 0.1 mm was used across the model. The effect of the mesh size was also considered in the study. Finally, the mesh size was selected so as not to generate long calculation times. Automatic remeshing was used to maintain a good mesh quality. Many various simulation methods, such as 2D axial–symmetric, 3D full simulation and 2.5D torsion approaches, were initially tested. Finally, a 2.5D torsion simulation including the circumferential effects was accepted for use in this study. A 2.5D model includes all the circumferential velocities needed to account for the torsional friction during welding. In the 2D axial–symmetric approach, the DEFORM code is not able to simulate a rotational speed, and the related shear deformation cannot be computed. The computing time required for a 3D full simulation is very long. In this paper, the torsional friction model has a significant impact not only on heat generation, but also on the deformation of the specimen, as noticed in reference [25]. Figure 2c shows the model with the sticking conditions at the contact point. This axisymmetric model takes into account both the axial force (see lower tooling) and the rotational speed (see upper tooling). The friction coefficient and surface pressures between the weld were taken into account. When the temperature is low, the friction stress is calculated using the Coulomb friction law according to the following formulae:

$$\tau_f = -\mu\sigma_n \frac{v_s}{|v_s|} \quad (1)$$

where τ_f is the shear friction, is the friction coefficient, σ_n is the normal stress, and v_s is the sliding velocity; the negative sign characterizes the opposition between the friction stress. The friction coefficient μ can be estimated as follows:

$$\mu = aP^bT^c exp(dV) \quad (2)$$

where P, T and V are the axial pressure, temperature and linear velocity, and a, b, c and d are the constants obtained from the experiments.

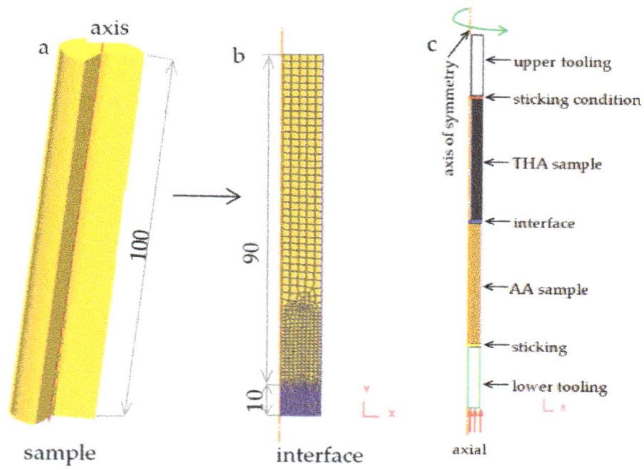

Figure 2. Two-dimensional axisymmetric model (**a**), mesh of the samples (**b**) and sticking condition (**c**).

When the temperature is high, the shear friction law is adopted and can be described as follows:

$$\tau_f = -mk\frac{v_s}{|v_s|} \quad (3)$$

where k is the shear yield stress of the material and m the friction factor [26].

In this study, to determine an appropriate shear friction, preliminary simulations were carried out. It was found that using a shear friction of 0.175 gave the best results for the given temperatures.

Data regarding the temperature-dependent thermal conductivity, specific heat and emissivity for AA were used; these were obtained from the DEFORM software library. The thermophysical properties of THA were taken from the ASM Handbook [2]. The properties of both materials for the simulation model are presented in Table 2.

The viscoplastic behavior of the materials is characterized by the temperature, flow stress and strain rate, and was obtained from the literature [2].

Figure 3 shows the flow stress for the aluminum (Figure 3a) alloy and tungsten composite (Figure 3b) at a strain of ε = 0.1 during deformation. The convective heat transfer coefficient was assumed to be 10 W/(m^2 K). The ambient temperature and initial temperature were set to 20 °C.

Table 2. Thermophysical properties of the materials.

Material	AA	THA
Density (kg/m^3)	2690	18,000
Young modulus (GPa)	69	385
Poisson ratio	0.33	0.28
Thermal expansion (1/C)	2.2×10^{-5}	5.2×10^{-6}
Specific heat (J/kg K)	528	134
Thermal conductivity (W/m K)	135	90
Melting point (°C)	640	2173

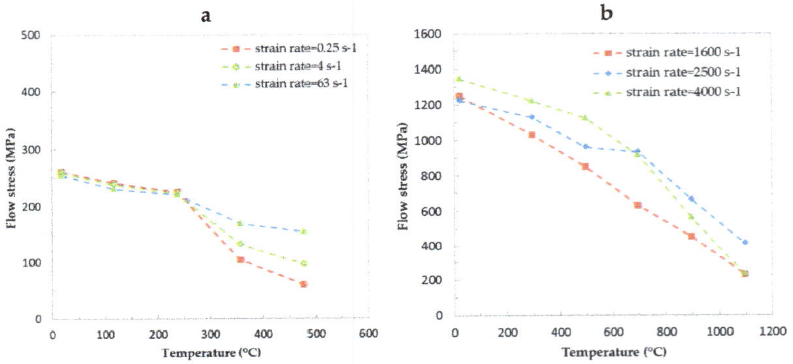

Figure 3. The flow stress of AA (**a**) and THA (**b**) at strain ε = 0.1 during deformation.

3. Results and Discussion

3.1. Temperature Distribution

The temperature field distributions during the friction stage of the FRW process of THA to AA are presented in Figure 4. As can be seen in the graphs, the friction welding process begins in the places most distant from the axis of the cylindrical samples. This is where the highest peripheral speed occurs, where the material becomes plasticized as a result of increasing temperature. Figure 4a shows a relatively rapid increase in temperatures during the friction period of the samples. After 3 s, the temperature on the surface reached 278 °C. After another 3 s, the temperature almost doubled (Figure 4b). It is likely that the high-temperature zone widens from the weld line due to heat conduction within the specimen. As can be seen in Figure 4c, no significant plastic deformation or flash formation occurred during the friction phase.

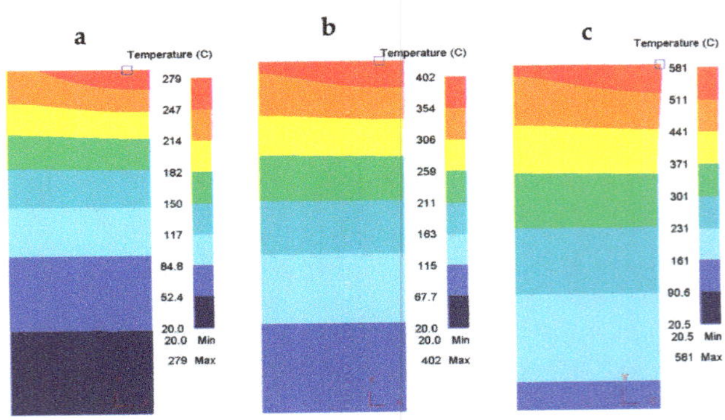

Figure 4. Distributions of temperature fields in aluminum alloy during the friction period: (**a**) 3 s (**b**) 6 s (**c**) 9.5 s. (FF = 12.5 kN, n = 1450 rpm, and UF = 50 kN).

The maximum interface temperature of 581 °C for the friction weld was achieved in a time of 9.6 s (Figure 5).

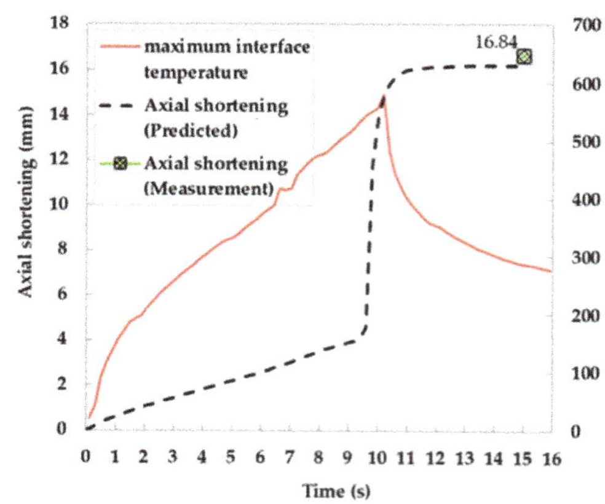

Figure 5. Maximum temperature distribution and axial shortening of welded materials during friction welding.

Figure 6 shows the temperature distribution fields of the weld zone during the upsetting stage of the welding process. As can be seen in Figure 5, after reaching the maximum value, the temperature drops rapidly to 280 °C within a few seconds during the forging stage (see Figure 6a–c).

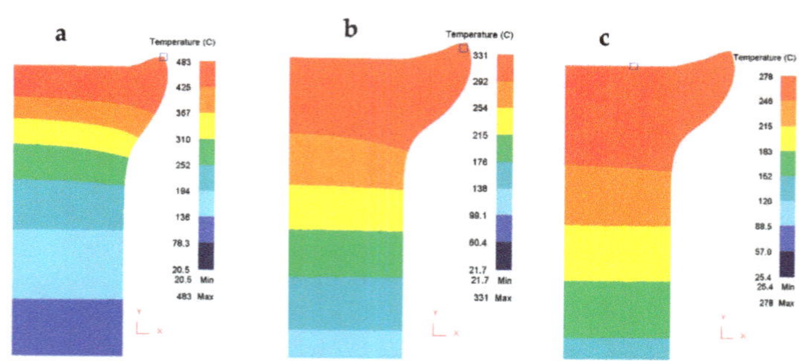

Figure 6. Temperature distribution of weld zone during the upsetting stage of the FRW process of THA/AA dissimilar alloys: (**a**) 11 s, (**b**) 13 s and (**c**) 15 s. (FF = 12.5 kN, n = 1450 rpm, and UF = 50 kN).

It is found that the weld flash begins from 9.6 s to 15 s and that the largest flash is obtained after the upsetting process (Figure 6c). The plastic deformation that accompanies this welding phase causes the significant shortening of the aluminium alloy (Figure 5). It can be seen that the maximum axial shortening of 16.8 mm is achieved during this stage.

Figure 7 shows the thermal history of the weld interface for various positions. It can be seen that the temperature on the outer surface is the highest, while that on the axis has the lowest value. The highest temperature values were recorded at the welding time of 9.5 s. For all measurement locations, the nature of the temperature distribution is similar. This is because the temperature increases linearly with the increasing friction time. After obtaining the highest value, the temperature drops rapidly, but on the outer surface, this reduction is the smallest.

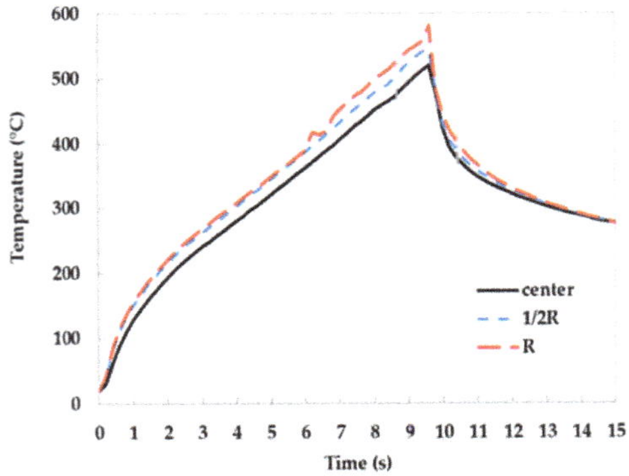

Figure 7. Thermal curves in the weld interface.

The radial distribution of temperature for various durations is obtained via simulation in Figure 8.

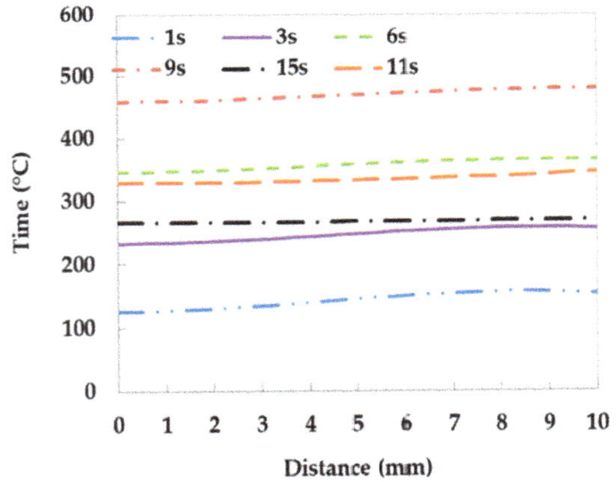

Figure 8. The radial temperature distribution at different times.

It can be noted that an appropriate welding time is required to reach the friction welding temperature of the welding process for specific parameters. The radial temperature was measured for the following welding parameters: rotational speed= 152 rad/s, friction force = 12.5 kN, upsetting time and force of 5 s and 50 kN, respectively. Similar results were obtained for other welding parameters.

3.2. Experimental Validation

3.2.1. Experimental Temperature Validation

The typical simulation result is obtained under a friction force, upsetting force, friction time, upsetting time and rotating speed of 12.5 kN, 50 kN, 9 s, 5 s and 1450 rpm, respectively. The computed and measured temperatures on the AA side are plotted for comparison. As shown in Figure 9, the thermocouples were placed in the middle of the radius and the axis of the joint at distances of 20, 25 and 30 mm from the welding line (see Figure 1b). Initially, the thermocouples were positioned close to the weld interface, but they were consumed during the forging phase of the welding process. Due to the continuous 'upset' (axial shortening) of the specimens during welding, it is naturally difficult to place thermocouples that can record exact temperatures. Consequently, the first thermocouple was located about 20 mm away from the weld interface. The thermocouple wires were stuck onto the surfaces of the measuring specimen with high-temperature-resistant glue.

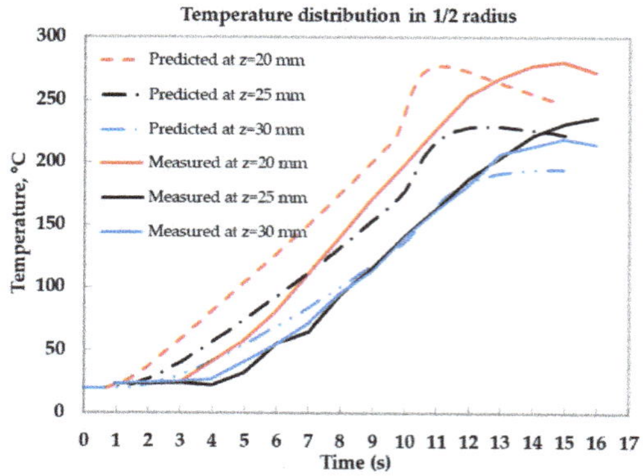

Figure 9. Predicted and measured temperature in AA side at the ½ radial position and at a distance of 0, 20, 25 and 30 mm from the original weld interface.

Figure 9 shows the similar characteristics of the temperature course during the heating of the materials. The predicted temperature values are slightly higher than the temperature values obtained from the thermocouple readings. According to the authors [27,28], these differences could be a result of reading the delay times and the time constant of the thermocouples.

Figure 10 shows the validation of the temperature at half of the sample radius.

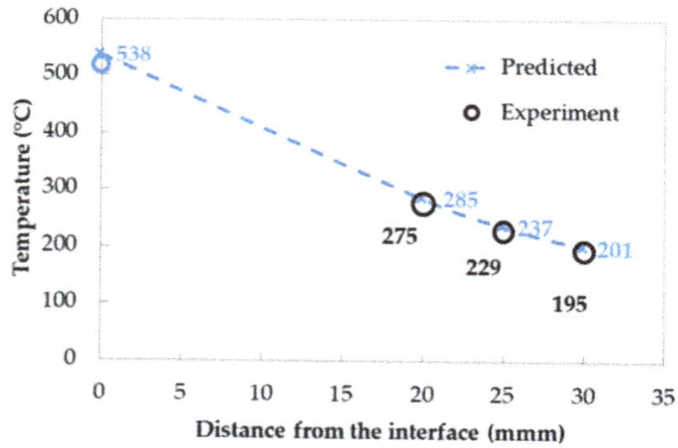

Figure 10. Validation of maximum temperature at the ½ radial position.

As can be seen from Figure 10, the predicted temperature is slightly higher than the temperature obtained by the experiment at distances of 20, 25 and 30 mm from bond line. The accuracy of the predicted temperature is quite good. The temperature results indicate small deviations of 3.6%, 3.5% and 3% for distances of 20, 25 and 30 mm from the interface. The maximum temperature near the welding point reached 538 °C in 1/2 of the sample radius for 9.5 s (Figure 10). Despite the plastic deformation of the material, this temperature did not reach the melting point of the aluminium alloy.

3.2.2. Macroscopic Morphology Validation

During the welding process, the plastic deformation of the joined materials occurs. The result of this process is the shortening of the soft metal and the formation of a weld flash. After completing the welding process, both the weld flash and the length of the samples were measured. A comparison of the macroscopic morphology of the friction welds using numerical simulation is shown in Figure 11.

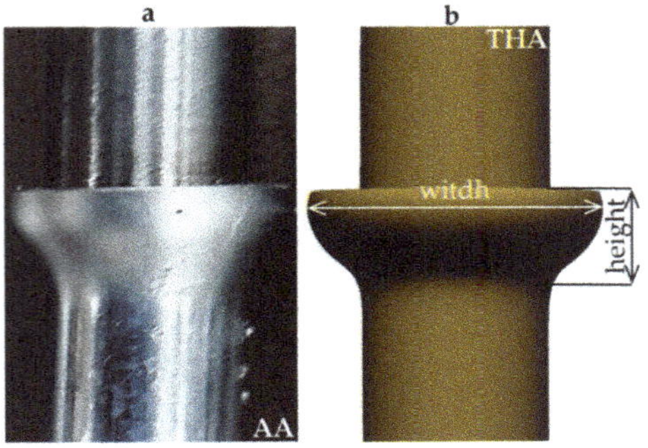

Figure 11. Comparison of the macroscopic morphology of the friction welds (**a**) and numerical simulation (**b**).

It can be seen from Figure 11 that the flash morphologies are nearly identical. A quantitative comparison of the flash dimensions obtained via the experiment and numerical simulations is shown in Table 3. Axial shortening is also selected for comparison.

Table 3. Comparison of flash dimension and axial shortening.

	Flash Dimensions		Shortening, mm
	Width, mm	Height, mm	
Measurement	33.66	16.84	16.61
Predicted	32.16	16.34	16.22

The results indicate very small deviations of 4.45%, 2.96%, and 2.34% in the flash width, flash height and axial shortening of the friction welds.

3.3. Effect of Friction Force on Interface Temperature

Figure 12 shows the influence of the frictional pressure force on the temperature in the joint during welding. The maximum temperature was obtained for the highest friction force. By increasing the friction force, the temperature rises and reaches the maximum interface temperature at a time of 7.5 s for all axial forces.

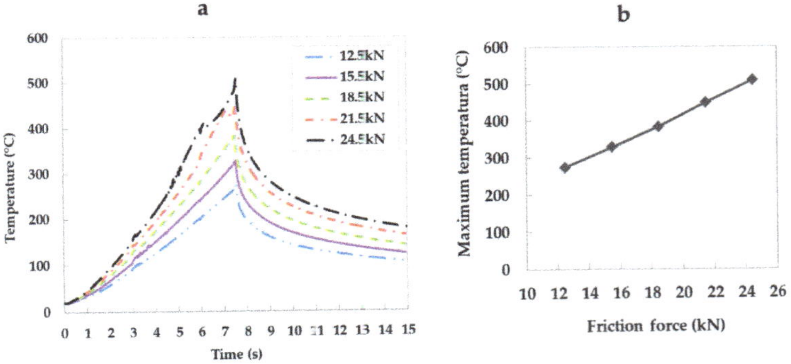

Figure 12. Influence of friction force on temperature during welding: temperature and various friction force in time (**a**), friction force and maximum temperature (**b**).

3.4. Effect of Friction Force on Axial Shortening

Figure 13 shows the effect of the friction force on axial shortening for friction forces of 12.5, 15.5, 18.5, 21.5 and 24.5 kN. It is clear that during the friction phase between materials, axial shortening takes place. As the friction time increases (from 12.5 to 18 kN), the axial shortening of the samples increases linearly. During the upsetting phase, it can be seen that a logarithmic shortening of the material takes place when using axial forces from 18.5 to 25 kN (Figure 13b).

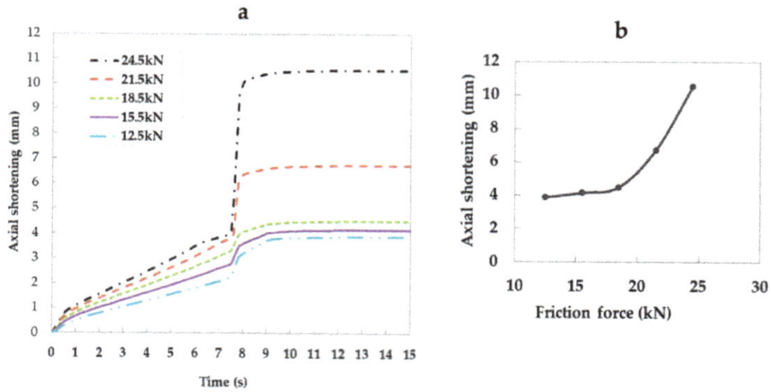

Figure 13. Effect of friction force on the axial shortening of the specimen: axial shortening and various friction force in time (**a**), friction force and axial shortening (**b**).

For all applied forces, the maximum shortening of the samples did not exceed 4 mm at the end of the friction phase. After this time, plastic deformation occurred and a weld flash was formed. This process took a few seconds. A rapid loss of material occurred after applying the upsetting force for a welding time of 8 s. The maximum material shortening was 3.5 mm for a friction force of 12 kN. Doubling the friction force shortened the material by an additional 7 mm (Figure 13b). Moreover, from Figure 13a, it can be seen that the forging period had a minimal effect on the upset. The axial shortening is almost unchanged throughout this stage. A similar observation reported in the references [29–31].

3.5. Effective Stress Distribution

Figure 14 shows the effective stress evolution at the three various positions of the weld interface.

As can be seen in Figure 14a, the effective stresses for the three positions during the friction period are relatively small. These stresses at the beginning of the friction process are 25, 45 and 50 MPa for the surface, axis, and 1/2 of the radius, respectively. The effective stresses on the sample surface are the smallest because no relative motion happens and only slight friction stress exists in this place [32]. The stresses gradually decrease and reach a value of 20 MPa at the end of the friction stage (Figure 14c) for all positions. At the beginning of the forging stage (Figure 14d), there is a sudden increase in stress, achieving 60 MPa and 80 MPa on the surface and in the axis of the joint, respectively. Constant effective stress takes place during forging from 10 to 12 s (Figure 14e). After stopping the welding process, the stress drops to zero. Moreover, it can be seen that the effective stress is lower in the periphery joint compared to the interior of the weld interface. According to the authors of reference [26], the friction stress is proportional to the shear yield strength.

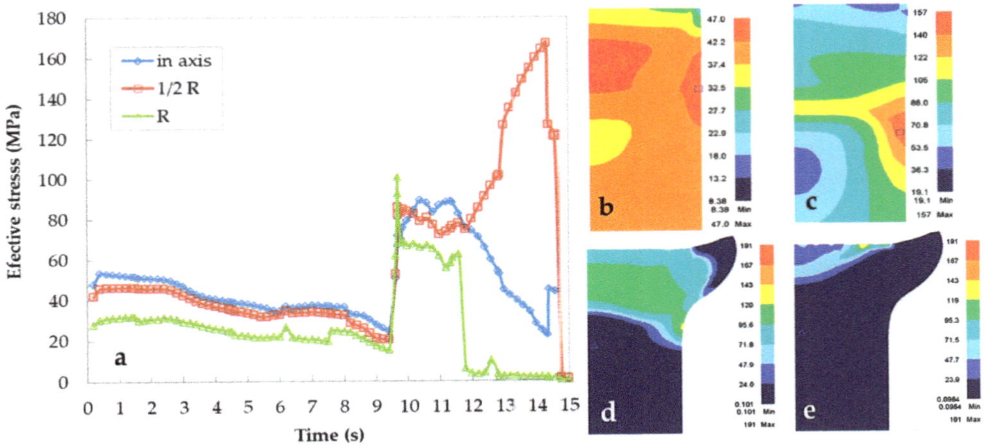

Figure 14. Predicted effective stress in the aluminum alloy specimen for various radial positions at the weld interface; effective stress flow during welding time (**a**), friction stage (**b**,**c**), forging stage (**d**,**e**).

4. Conclusions

An experimental and FEM simulation study of the CDFW process of a THA with an aluminum alloy is presented. The temperature distributions, effective stress, flash dimensions and axial shortening are calculated on un-rotated friction-welded aluminium parts. An increase in the friction force causes an increase in temperature during the friction period of the samples. It can predict the shape of the flash with a high degree of accuracy. The predicted temperature was slightly higher than the temperature obtained in the experiment. The maximum interface temperature of 581 °C is below the melting temperature of the aluminum sample. The results indicate very small deviations of 4.45%, 2.96%, and 2.34% in the flash width, flash height and axial shortening of the friction welds. With an increasing friction force, axial shortening increases exponentially. The effective stresses on the sample surface are the smallest because no relative motion occurs and only slight friction stress exists in this place.

Author Contributions: Conceptualization and Methodology, R.W. and A.S.; Software, R.W.; Modelling and optimization, R.W.; Validation, R.W. and A.S.; Formal analysis, R.W.; Investigation, R.W., A.S. and P.S.; Resources, R.W.; Writing-original draft preparation, R.W.; Writing-review and editing, R.W.; Visualization, R.W.; Supervision, R.W. All authors have read and agreed to the published version of the manuscript.

Funding: This research received no external funding.

Institutional Review Board Statement: Not applicable.

Informed Consent Statement: Not applicable.

Data Availability Statement: Data are contained within the article.

Acknowledgments: The authors wish to thank Mieczysław Kaczorowski from the Faculty of Mechanics of Institute of Mechanics and Design, Warsaw University of Technology, for his valuable suggestions.

Conflicts of Interest: The authors declare no conflicts of interest.

References

1. American Welding Society. Specifications and standards. In *Recommended Practice for Friction Welding*; American Welding Society: Miami, FL, USA, 1989.
2. Withers, P.J.; Preuss, M. *Simulation of Rotational Welding Operations*; Metals Process Simulation, ASM Handbook; ASM International: Materials Park, OH, USA, 2010; Volume 22B.

3. Vill, V.I. *Friction Welding of Metals*; AWS: New York, NY, USA, 1962.
4. Rich, T.; Roberts, R. Thermal Analysis for Basic Friction Welding. *Met. Constr. Br. Weld. J.* **1971**, 3, 93–98.
5. Imshennik, K.P. Heating in Friction Welding. *Weld. Prod.* **1973**, 20, 76–79.
6. Cheng, C.J. Transient temperature distribution during friction welding of two similar materials in tubular form. *Weld. J.* **1963**, 12, 223–240.
7. Healy, J.; Mcmullan, D.J.; Bahrani, A.S. Analysis of Frictional Phenomena in Friction Welding of Mild Steel. *Wear* **1976**, 37, 279–289. [CrossRef]
8. Kinley, W. Inertia Welding: Simple in Principle and Application. *Weld. Met. Fabr.* **1979**, 585–589.
9. Murti, K.; Sundaresan, G.; Parameter, S. Optimization in Friction Welding Dissimilar Materials. *Met. Constr. Br. Weld. J.* **1983**, 15, 331–335.
10. Kleiber, M.; Służalec, A. Finite element analysis of heat flow in friction welding. *Eng. Trans.* **1984**, 32, 107–113.
11. Służalec, A. Thermal Effects in Friction Welding. *Int. J. Mech. Sci.* **1990**, 32, 467–478. [CrossRef]
12. Służalec, A. Solution of thermal problems in friction welding. *Int. J. Heat Mass Transf.* **1993**, 36, 1583–1587. [CrossRef]
13. Sahin, Z.; Yilbas, B.S.; Al-Garni, A.Z. Friction welding of Al–Al, Al–steel, and steel–steel samples. *J. Mater. Eng. Perform.* **1996**, 5, 89–99. [CrossRef]
14. Bendzsak, G.J.; North, T.H.; Li, Z. Numerical Model for Steady-State Flow in Friction Welding. *Acta Mater.* **1997**, 45, 1735–1745. [CrossRef]
15. Fu, L.; Duan, L. The Coupled Deformation and Heat Flow Analysis by Finite Element Method during Friction Welding. *Weld. Res. Suppl.* **1998**, 77, 202–207.
16. Alvise, L.D.; Massoni, E.; Walloe, S.J. Finite Element Modelling of the Inertia Friction Welding Process between Dissimilar Materials. *J. Mater. Process. Technol.* **2002**, 125, 387–391.
17. Balasubramanian, V.; Li, Y.; Stotler, T.; Crompton, J.; Soboyejo, A.; Katsube, N.; Soboyejo, W.A. New Friction Law for the Modelling of Continuous Drive Friction Welding: Applications to 1045 Steel Welds. *Mater. Manuf. Process.* **1999**, 14, 845–860. [CrossRef]
18. Zimmerman, J.; Włosiński, W.; Lindemann, Z.R. Thermo-mechanical and diffusion modelling in the process of ceramic-metal friction welding. *J. Mater. Process. Technol.* **2009**, 209, 1644–1653. [CrossRef]
19. Ambroziak, A. Temperature distribution in friction welded joints of dissimilar refractory metals. *Adv. Manuf. Sci. Technol.* **2002**, 26, 39–54.
20. Maalekian, M.; Kozeschnik, E.; Brantner, H.P.; Cerjak, H. Comparative analysis of heat generation in friction welding of steel bars. *Acta Mater.* **2008**, 56, 2843–2855. [CrossRef]
21. Xiong, J.T.; Li, J.L.; Wei, Y.N.; Zhang, F.S.; Huang, W.D. An analytical model of steady-state continuous drive friction welding. *Acta Mater.* **2013**, 61, 1662–1675. [CrossRef]
22. Winiczenko, R.; Goroch, O.; Krzyńska, A.; Kaczorowski, M. Friction welding of tungsten heavy alloy with aluminium alloy. *J. Mater. Process. Technol.* **2017**, 246, 42–55. [CrossRef]
23. Winiczenko, R.; Kaczorowski, M.; Krzyńska, A.; Goroch, O.; Skibicki, A.; Skoczylas, P. TEM Microstructure, Mechanical Properties and Temperature Estimation in the 5XXX Series Al-Mg-Si Aluminum Alloy with W-Ni-Fe Tungsten Composite Friction-Welded Joints. *Materials* **2022**, 15, 1162. [CrossRef]
24. Perovic, N.L.; Maglic, K.D.; Stanimirovic, A.M.; Vukovic, G.S. Transport and calorimetric properties of AISI 321 by pulse thermal diffusivity and calorimetric techniques. *High Temp. High Press* **1995**, 1, 53–58. [CrossRef]
25. Hazman, S.; Ismail, A.I.M.; Endri, R.; Zainal, A.H. Mechanical evaluation and thermal modeling of friction welding of mild steel and aluminium. *J. Mater. Process. Technol.* **2010**, 210, 1209–1216.
26. Zang, Q.Z.; Zhang, L.W.; Liu, W.W.; Zhang, V.G.; Qu, S. 3D rigid viscoplastic FE modelling of continuous drive friction welding process. *Sci. Technol. Weld. Join.* **2006**, 11, 737–743. [CrossRef]
27. Shailesh, K.; Singh, K.; Chattopadhyay, G.; Phanikumar, K.; Dutta, P. Experimental and numerical studies on friction welding of thixocast A356 aluminum alloy. *Acta Mater.* **2014**, 73, 177–185.
28. Le, W.S.; Xiea, G.L.; Lin, C.F. The strain rate and temperature dependence of the dynamic impact response of tungsten composite. *Mater. Sci. Eng.* **1998**, 257, 256–267. [CrossRef]
29. Li, W.; Wang, F. Modeling of continuous drive friction welding of mild steel. *Mater. Sci. Eng. A.* **2011**, 528, 5921–5926. [CrossRef]
30. Geng, P.; Qin, G.; Zhou, J. Numerical and experimental investigation on friction welding of austenite stainless steel and middle carbon steel. *J. Manuf. Process.* **2019**, 47, 83–97. [CrossRef]
31. Wang, F.F.; Li, W.Y.; Li, J.L.; Vairis, A. Process parameter analysis of inertia friction welding nickel-based superalloy. *Int. J. Adv. Manuf. Technol.* **2014**, 71, 1909–1918. [CrossRef]
32. Nu, H.; Le, T.; Minh, L.; Loc, N. A Study on Rotary Friction Welding of Titanium Alloy (Ti6Al4V). *Adv. Mater. Sci. Eng.* **2019**, 2019, 4728213.

applied sciences

MDPI

Article

The Effect of Tool Rotation Speed on the Formation of Eutectic Structure during Friction Stir Welding of Aluminum to Magnesium

Kiarash Torabi [1], Reza Beygi [1,2,*], Ghasem Eisaabadi Bozchaloei [1] and Lucas F. M. da Silva [3]

1 Department of Materials Engineering and Metallurgy, Faculty of Engineering, Arak University, Arak 3815688349, Iran; kiarash.torabi@hotmail.com (K.T.)
2 Institute of Science and Innovation in Mechanical and Industrial Engineering (INEGI), Rua Dr. Roberto Frias, 4200-465 Porto, Portugal
3 Department of Mechanical Engineering, Faculty of Engineering of the University of Porto, Rua Dr. Roberto Frias, 4200-465 Porto, Portugal
* Correspondence: rbeygi@inegi.up.pt

Abstract: Friction stir welding (FSW) is a solid-state welding process capable of joining a wide range of light metals. However, liquation and solidification may occur during joining of dissimilar metals which leads to eutectic formation. This article aims to discover the influence of tool rotation speed on the formation of eutectic structure during friction stir welding of aluminum to magnesium. To do so, friction stir welding was performed at 600 and 950 rpm to join pure aluminum and ECO-AZ91 magnesium alloy in a lap configuration. In order to investigate the influence of the welding speed, the welding speeds of 23.5 and 37.5 mm/min were also chosen. Scanning electron microscopy (SEM) was used to study the microstructure of the joints. A shear-tensile test was used to evaluate the joints' strengths. The fracture surfaces were also studied by SEM. The results revealed that changing the rotation speed directly affects the eutectic formation, whereas the welding speed had no influence. A lower rotation speed resulted in a thin, continuous intermetallic layer, whereas a higher speed led to the formation of a massive Mg-$Al_{12}Mg_{17}$ eutectic microstructure. The formation of eutectic, as an indicative of liquation, may affect the material flow during the process due to decreasing the friction coefficient between the tool and material. The macrostructure analyses showed that the phase evolution as well as the mechanism of material flow are highly affected by liquation.

Keywords: FSW; aluminum to magnesium joining; intermetallic compounds; eutectic structure; liquation

check for updates

Citation: Torabi, K.; Beygi, R.; Eisaabadi Bozchaloei, G.; da Silva, L.F.M. The Effect of Tool Rotation Speed on the Formation of Eutectic Structure during Friction Stir Welding of Aluminum to Magnesium. *Appl. Sci.* **2023**, *13*, 7133. https://doi.org/10.3390/app13127133

Academic Editors: Ivan Galvão, Rui Manuel Leal and Gustavo H. S. F. L. Carvalho

Received: 18 May 2023
Revised: 8 June 2023
Accepted: 12 June 2023
Published: 14 June 2023

1. Introduction

The growth in engineering in industrial applications has led to efforts to construct cost-efficient alloys with exceptional properties, such as high strength, light weight, and ductility [1]. Despite their attractive characteristics and widespread usage, aluminum (Al) alloys require further development [2]. In this regard, confronting this obstacle is achieved by constructing Al-based alloys, which have garnered attention due to their light weight, higher strength, and toughness, as well as reusability [3]. Though other materials' properties need to be considered as corrosion resistance [4,5], many studies tend to focus on mechanical properties [6]. Aluminum and its alloys are widely used in industry due to their low density, high specific strength, good corrosion resistance, good performance, high thermal and electrical conductivity, attractive appearance, and recyclability [7–9]. As a very light metal, magnesium and its alloys have excellent specific strength, good castability, hot formability, good shielding against electromagnetic interference, and recyclability [10]. To save weight, aluminum alloys, and magnesium alloys can be used instead of steel and iron, which are much heavier [10,11]. There are specific programs to use more suitable metals instead of other metals due to the advantages of their specific characteristics, such as the damping ability of magnesium and the creep resistance of aluminum [9,12]. It is expected that both metals, aluminum, and magnesium, will be used more for industry in the future. The combination of aluminum and magnesium in a composite structure allows the use of these alloys for more applications, which results in optimal weight savings.

Among others, a dissimilar joint of Al and magnesium (Mg), labeled Al/Mg, has been found as a promising approach for reaching a high strength-to-weight ratio property [13]. Having various structural and mechanical properties makes the joining of these two alloys attractive due to the possibility of the formation of wide properties, which were inaccessible before [14].

There is quite a vast range of methods utilized for the dissimilar welding of Al/Mg; thus, choosing the joint method is of paramount importance. Considering the drawbacks of Al, namely low molten viscosity, high reflectivity, and inherent oxide layer formation, traditional welding processing resulted in hot cracking, poor joining, and potential defects [7]. Correspondingly, the formation of intermetallic compounds, brittle compounds, pores, and solidification cracks in conventional approaches, e.g., arc welding and fusion welding processes, was observed [15]. Compared to others, to overcome these shortcomings, friction stir welding (FSW) has been suggested because of its processing uniqueness and cost-efficient technique [16,17]. It is worth mentioning that FSW prevents the formation of weld defects and reduces hydrogen solubility [18]. During FSW, the welding tool contains two different motions: rotary and linear along the joint line. The pin is plunging between the sheet workpieces with rotational motion followed by traversing along the joint causing dissimilar bonding by mixing metals at the joint zone [9]. The intermixing of two materials at the stir zone is affected by engaged processing parameters such as welding speed, tool rotation speed, and the piece position [19–21]. In this context, it is figured out that applying optimized rotational and linear speed causes the formation of stable phases in the weld zone [22].

During the welding process, the kinetic energy is turned into thermal energy due to friction at the interface between the tool and work pieces; thus, process parameters strictly impact the amount of introduced heat [9]. The tool traverse speed and rotational speed remarkably affect the heat generation. It is recommended that the heat input is directly and inversely proportional to rotational speed and tool traverse speed, respectively [7]. As is well documented, increasing rotational speed eventuated in higher heat input during FSW [23–26]. Therefore, to gain higher ultimate tensile strength (UTS), the heat input could be controlled by tool traverse and rotational speed [27]. However, sufficient heat input results in recrystallization through the weld zone leading to the formation desired microstructure [9]. In addition to the influence on heat input, tool rotational speed affects the fracture position of the weld, i.e., increasing rotational speed not only resulted in broadening the strained zone, but also the position of the maximum strain region moves from the retreating side (RS) to the advancing side (AS) of the weld [28]. The intermixing leads to the formation of intermetallic compounds (IMCs) in the stir zone which detrimentally affects joint quality, ductility, and the mechanical properties of the joint [29,30]. Based on various studies, IMCs are able located in three positions: (1) coherent IMCs layer at the interface, (2) IMCs fragments in Mg matrix, and (3) IMCs fragments in the Al matrix [31]. Due to enhanced diffusion during welding, although the restriction of formation and growth in IMCs are not accessible, reducing the thickness of IMCs is one way to enhance the weld strength which could be achieved by reducing the frictional heat input [32,33]. The formation of IMCs depends on heat input and local composition, regardless of the chemical composition of bare material. The IMCs formation at the interface of Al and Mg in FSW joints is attributed to diffusion or eutectic reaction mechanisms [34]. In addition to welding parameters, heat is governed by plastic deformation (E_d), friction (E_f) at the tool/workpiece interface, and viscous dissipation (E_v) which is controlled by a difference in friction coefficient, liquation susceptibility, and deformability of Al and Mg [14]. For instance, the eutectic reaction ($Mg + Al_{12}Mg_{17} \rightarrow L$) resulted in the formation of liquid films causing reduced resistance to tool rotation due to tool slippage and interfering with plastic deformation [35]; consequently, due to declining E_d, E_f, and E_v, heat generation is significantly decreased. Deformability strictly depends on crystal structure, e.g., in the case of Al/Mg, Al possesses a face-centered cubic (fcc) structure whereas Mg has a hexagonal close-packed (hcp) structure offering 12 and 3 slip systems, respectively. As a result, Al deformability is much better than Mg, and heat is generated through E_d and E_v [34].

Considering that the FSW process temperature remains below eutectic temperatures; therefore, the conclusion can be made that the formation of IMCs at the interface is associated with diffusion between Al and Mg atoms [36]. On the other side, despite the fact that the welding temperature is lower than the eutectic temperature, the liquation/melting due to heat produced by the rotation tool forms a thin liquid film along grain boundaries leading to growth in IMCs [37]. It is worth mentioning that the high rotational speed may cause defects formation and imperfect joining at the interface, resulting in lower tensile strength [38]; as a result, tensile strength is affected by thermal history (i.e., heat input and welding time). Considering the aforementioned approaches, maintaining the temperature below the eutectic temperature is a practical way to restrict the formation of IMCs or reduce IMCs thickness. The weld strength is mainly influenced by two phenomena: the existence and distribution of IMCs as well as the material interlocking phenomenon [39,40]. The interlocking of materials at the interface is one way to strengthen welding; to be more accurate, since the heat input is maintained below the eutectic temperature to control IMCs formation, weld quality could improve by mechanical interlocking [41].

In the quest for higher welding strength, the influence of the involved parameters, mentioned above, and tool geometry have been elucidated which affirms that the formation of defects is highly related to rotational speed and tool geometry [42]. The influence of tool traverse speed on mechanical properties shows that the maximum UTS of 147 MPa with a joint efficiency of 61% was achieved with optimum parameters [43]. The work-piece position of materials plays a pivotal role in welding quality, i.e., the FSW tool contains two important parts: pin and shoulder; as a result, welding strength simply varies by which material is connected with the shoulder [44]. It is reported that the tensile strength is tied up to sheet position and pin penetration, which is ascribed to different IMCs formation at the weld zone [45,46]. Ghiasvand et al. studied the influence of double-welding on mechanical properties which results in a higher tensile-shear force. They observed two different fracture morphologies, i.e., brittle and ductile, due to the formation of IMCs [47]. Furthermore, it is reported that variations in shoulder diameter, pin height, and pin diameter affect mechanical properties which is due to changes that occurred in the contact area of the welding tool and workpiece resulting in diverse heating input [48]. Accordingly, the reaction that occurred at the interface of base metal is influenced by the pin profile and welding parameters resulting in formation of thick brittle IMCs which are a combination of Al_3Mg_2 and $Al_{12}Mg_{17}$ [49]. Furthermore, the cladding Al to Mg was observed when an Al layer was on top of Mg, which is a significant benefit [50].

As a part of our ongoing welding of Al to Mg by the FSW method; herein, we have investigated the effect of intermixing of Al and Mg during FSW and its effect on the weld failure in respect of varying processing parameters, namely welding traverse speed and tool rotation speed. The used configuration was in a way that Al was placed on the top of Mg. To elucidate the influence of the FSW parameters on eutectic formation, a work window of the welding parameters was selected as a way to achieve various welding intermixing in the stir zone.

2. Materials and Methods

Here, aluminum AA1050 and AZ91 magnesium with thicknesses of 3 mm were used and the chemical composition of the bare sheets are given in Tables 1 and 2. To remove surface oxide layers and contaminations, the AA1050 and AZ91 sheets were cleaned with acetone followed by scratch wire-brushing. The sheets were placed in an AA1050/AZ91 lap joint configuration and the welding process was carried out with different welding traverse speeds and tool rotation speeds; the whole process is depicted in Figure 1. The used tool was made of an H13 tool steel containing the shoulder with diameter of 18 mm and the pin diameter of 5 mm and length of 4.7 mm. The tool tilt angle was maintained at 2.5° for the tool during the joining process and the formation of welds. At the constant inclination angle, to investigate the influence of tool traverse and rotational speed, the samples with different tool rotation speeds and the traverse speeds were fabricated and listed in Table 3. In the welding process, as schematized in Figure 1, the pin was penetrated into the bottom layer (AZ91).

Table 1. Chemical composition of AA1050 (wt%).

Al	V	Zn	Mn	Fe	Si	Cu
Balance	0.010	0.050	0.050	0.42	0.2	0.08

Table 2. Chemical composition of AZ91 (wt%).

Mg	Fe	Si	Ni	Cu	Mn	Zn	Al	Ca
Balance	0.005	0.3	0.005	0.051	0.202	0.403	8.14	0.577

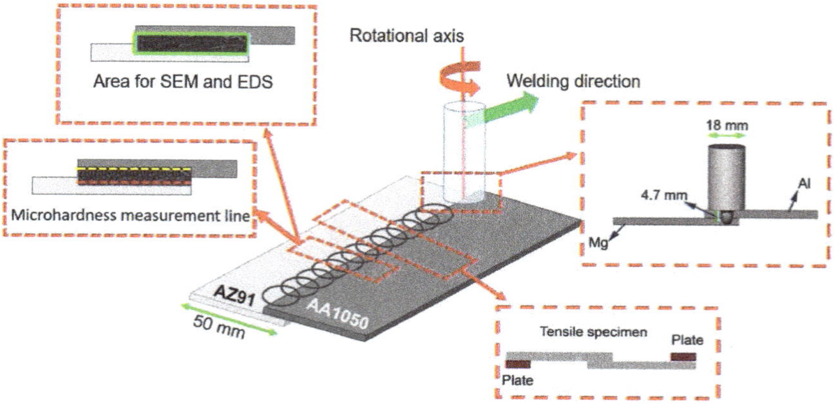

Figure 1. An illustration of the welding process representing the geometry of the tool and the location of the samples for the performed tests. The lap configuration was Al/Mg and the tool was inserted into Al. The advancing side was placed in aluminum. The length of the weld was 100 mm.

Table 3. The welding parameters used for friction stir lap welding of AA1050/AZ91.

Sample	Welding Speed (mm/min)	Tool Rotation Speed (rpm)
A	23.5	950
B	37.5	950
C	23.5	600
D	37.5	600

To visualize the morphology and chemical analysis of samples, scanning electron microscopy (SEM, model: Mira2 TESCAN) equipped with an energy-dispersive X-ray spectrometer (EDS) was used. To this end, samples were prepared using a standard metallographic procedure, which involved cutting, grinding, and polishing the samples to a mirror finish. In order to investigate shear-tensile test, different specimens were prepared for, as depicted in Figure 1.

To prepare samples for tensile experiments, ASTM D3164 was considered as a guideline. Shear-tensile tests were carried out with a constant crosshead velocity of 2 mm/min. As shown in Figure 1, during the shear-tensile test, to adjust the offset in the lap joint, spacer sheets were used. To avoid experimental errors and ensure repeatability, three shear-tensile specimens were examined. The offset of lap joint was achieved by using spacer sheets located between the clamps and the specimen ends, as seen in Figure 1.

The Vickers hardness test method was utilized, which is a widely accepted method for measuring the hardness of metallic materials. This method involves applying a load to the specimen using a diamond indenter and measuring the size of the resulting impression to calculate the hardness. The microhardness measurement line for samples in AA1050 and AZ91 sides is depicted in Figure 1.

3. Results and Discussion

To visualize the welding appearance of samples, the top views of the joint zone for samples (A–D) are shown in Figure 2. Throughout Figure 2, no serious macro defects such as lack of connection, grooves, flashes, and warm defects appeared. Moreover, in the case of samples with rotation speed of 950 rpm, a small material outflow was observed. The cross-sections of the joints are presented in Figure 3 and the advancing side (AS) and retreating side (RS) are labeled. In all samples, AA1050 was dragged into the AZ91 layer at the center of the weld. In some samples, a hook can be observed in the retreating side in which AZ91 was dragged into the AA1050 layer.

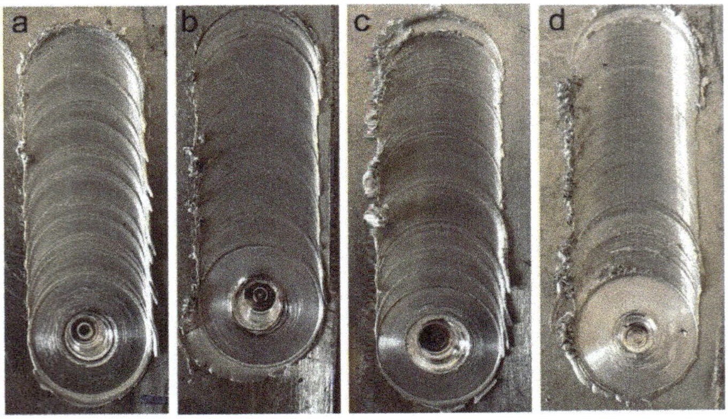

Figure 2. (**a–d**) Top views of the welds zone of AA1050/AZ91 joints for A–D samples, respectively. (Welding condition was introduced in Table 3).

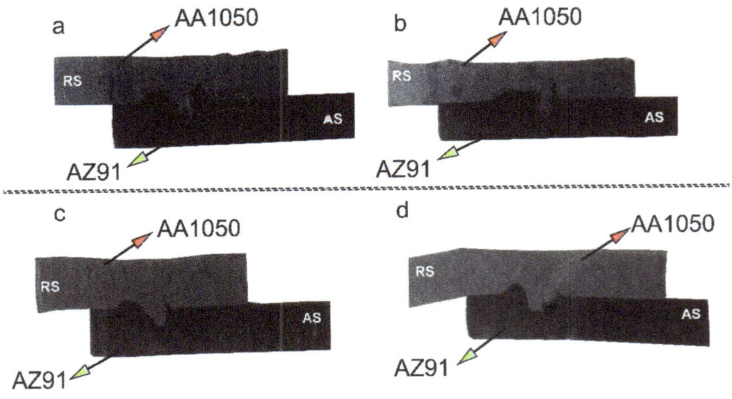

Figure 3. (**a–d**) The cross-sections of the joints for A–D samples, respectively. (Welding condition was introduced in Table 3).

It should be noted that one way to understand the chemical bonding mechanism between AA1050/AZ91 is their properties and incompatibility aspect. Generally, an impractical feature is a limited solubility governing the formation of IMCs which is explained by the binary phase diagram for these elements, see Figure 4 adapted from Ref. [7]. Nevertheless, the welding process is non-equilibrium, and to predicate the formation phases and possible reactions during welding, the equilibrium binary phase diagram is a standard guideline. Considering the fact that measuring the exact peak temperature during welding is almost impossible, using phase diagram is an acceptable method to envisage formation IMCs which depends on local temperature and composition. Throughout Figure 4, several phenomena can be comprehended: the binary system includes two eutectic lines at 437 and 450 °C which are lower than melting points of pure AA1050 and AZ91. Furthermore, three stable intermetallic phases including Al_3Mg_2 (complex cubic β) phase, $Al_{12}Mg_{17}$ (cubic of α-Mg type γ) phase and rhombohedral R phase exist [7,34].

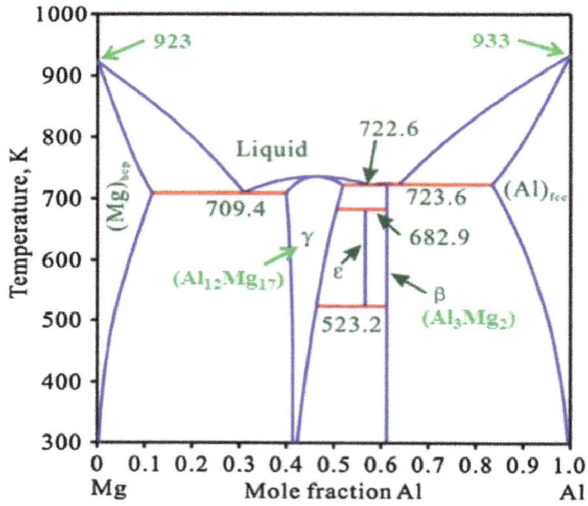

Figure 4. Equilibrium phase diagram of Al-Mg system [7].

As stated, involved parameters have pivotal role on the formation of microstructure. In order to visualize heat input's influence on the fracture behavior of the welds and find a correlation between welding parameters and joint strengths, the cross sections of the welds were studied by SEM in Figure 5. The SEM image of the weld nugget zone of sample A is shown in Figure 5a. A continuous layer of intermetallic compounds is not observed at the interface of AA1050 and AZ91. In Figure 5a, an SEM image of the interface at a higher magnification reveals the presence of a liquefaction region (highlighted in pink). This region appears to consist of two distinct phases. The SEM image of sample B is shown in Figure 5b at different magnifications revealing that the lamellar eutectic structures consisting of Mg + $Al_{12}Mg_{17}$ were formed at the AA1050/AZ91 interface. At the interface of AA1050 and AZ91, a region was observed which consists of a continuous bilayer IMC and a liquefaction layer. This is consistent with another report in which the eutectic layer was formed in the Mg-rich region [51]. It should be noted that the liquefaction leads to eutectic formation in the stir zone which possesses a different formation mechanism compared to IMCs [7]. The SEM interface of sample C is shown in Figure 5c, and a very thin continuous layer at the interface of AA1050 and AZ91 was observed. Figure 5d shows the SEM image of sample D, and a very thin continuous layer can be observed at the interface of AA1050 and AZ91. Comparing SEM images revealed that in sample D, a small cavity on the AS, which is colored in the orange region in the nugget zone, emerged. This is probably due to lower heat input resulting in insufficient mixing and plasticization of the materials.

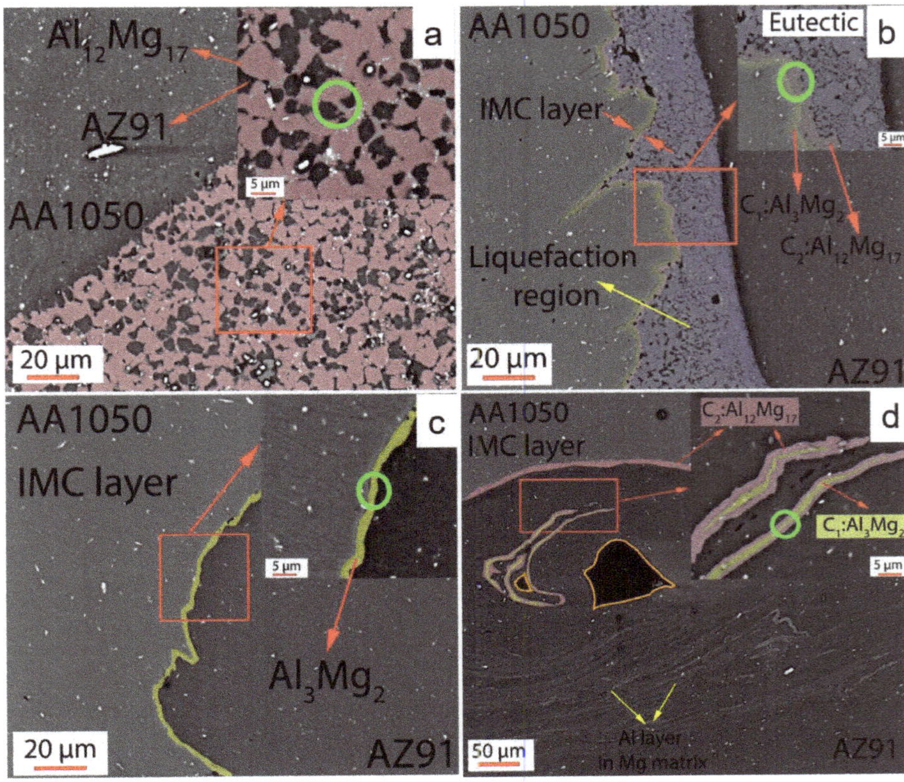

Figure 5. The SEM images of (**a–d**) sample A–D, respectively. The continuous layer, liquefaction region, and eutectic structure were marked in the panels. In panel (**d**), a formed cavity is highlighted in orange.

To study the chemical composition of samples, the EDS analysis of these regions shown in Figure 6 was performed from places indicated by the green circle in Figure 5. The EDS spectrum of sample A was depicted in Figure 6a, and the results suggest that this pink region is composed of a Mg–$Al_{12}Mg_{17}$ eutectic compound. The EDS analysis of the continuous IMC layer of sample B in Figure 6b shows that the liquefaction zone formed a eutectic phase which is composed of Mg and $Al_{12}Mg_{17}$ and a continuous layer consisting of $Al_{12}Mg_{17}$ and Al_3Mg_2 was formed adjacent to the Al layer. The EDS spectrum of sample c, as shown in Figure 6c, provides confirmation that at the interface of AA1050 and AZ91, the presence of Al_3Mg_2 has been detected, indicating the formation of a continuous layer. The EDS analysis of sample D, Figure 6d, revealed that $Al_{12}Mg_{17}$ was formed at the interface of AA1050 and AZ91. Moreover, fragments of AA1050-AZ91 intermetallic phases are dispersed in AZ91 matrix, whereas in AA1050 matrix dispersion of AA1050-AZ91, intermetallic phases cannot be detected.

The formation of a brittle intermetallic compound (IMC) layer at the boundary between the base metals is the most critical issue that limits the strength of the dissimilar FSW joint between AA1050 and AZ91. The tensile test results, as a function of displacement, are presented in Figure 7a and the highest obtained UTS for samples is presented in Figure 7b. An illustration of the fracture path was schematized in Figure 7c. A great difference can be observed in some samples indicating that welding parameters influence the fracture strength. The fracture samples were precisely examined, and the fracture area was schematically shown in Figure 7c for all samples. It is noticeable that the fracture for sample A took place across the eutectic zone. Similar to sample A, the fracture in sample B was placed in the eutectic zone as shown schematically in Figure 7c. Despite the presence of a continuous layer of intermetallic compounds, the fracture did not occur through this layer due to a low thickness of this continuous IMC layer. In this sample, the liquation during welding caused a decrease in the friction coefficient which in turn decreased the heat generation during welding [34]. Furthermore, the formation of a eutectic microstructure impeded the growth of a thick IMC layer due to hindering diffusion and activity of elements which is consistent with another study [37]. Thus, it could be concluded that a lower local temperature during welding and eutectic formation leads to the formation of a thinner intermetallic compound at the interface. The illustration of the fracture place for sample C demonstrates that the fracture took place at the interface of Al_3Mg_2 and Mg. According to Figure 7c, for sample D, it can be seen that the fracture was propagated at the interface of $Al_{12}Mg_{17}$ and Mg. Some brittle regions at the fracture surface can be observed which is due to the dispersion of intermetallic phases in Mg matrix as seen in Figure 5d.

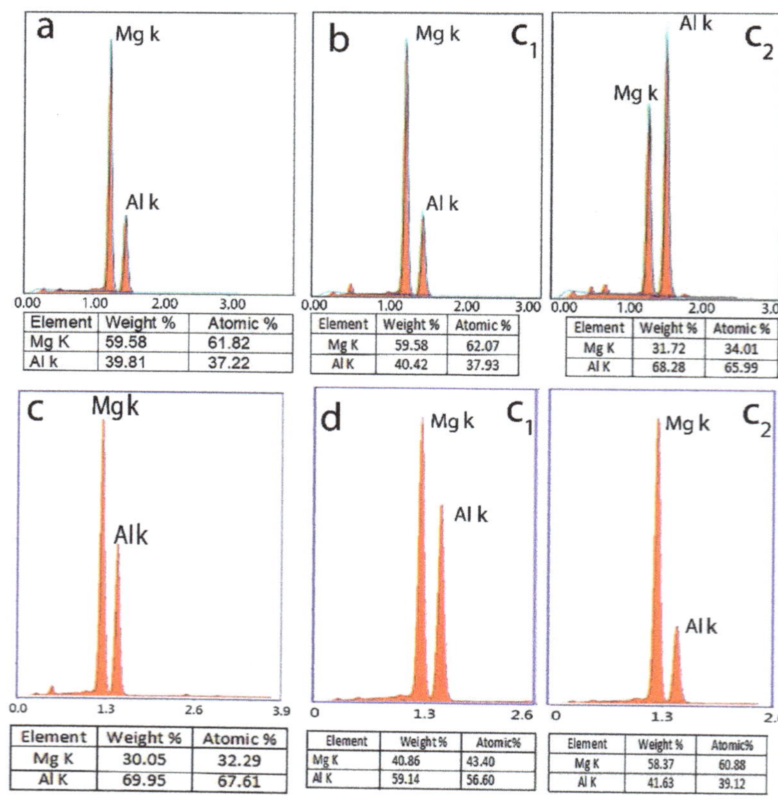

(a)

Element	Weight %	Atomic %
Mg K	59.58	61.82
Al k	39.81	37.22

(b) C1

Element	Weight %	Atomic %
Mg K	59.58	62.07
Al K	40.42	37.93

(b) C2

Element	Weight %	Atomic %
Mg K	31.72	34.01
Al K	68.28	65.99

(c)

Element	Weight %	Atomic %
Mg K	30.05	32.29
Al K	69.95	67.61

(d) C1

Element	Weight%	Atomic%
Mg K	40.86	43.40
Al K	59.14	56.60

(d) C2

Element	Weight %	Atomic%
Mg K	58.37	60.88
Al K	41.63	39.12

Figure 6. (**a**–**d**) The EDS spectrum of A–D, respectively. The C_1 and C_2 in panel (**b**,**d**) represent the EDS of regions shown in Figure 5b and 5d, respectively.

It is documented that the fracture location (Figure 7c) occurred through the bonding interface which is mainly attributed to presence of brittle IMCs at the Al/Mg interface, adversely affecting the joint quality. Meticulous investigation of tensile test of samples (Figure 7a) revealed that three types of trends can be derived: (1) when tool rotational speed is steady (i.e., samples A and B), a sample with lower welding traverse speed (sample A) shows a higher, ultimate tensile strength (UTS); (2) at constant welding traverse speed (i.e., samples A and C), the lower rotational speed (sample C) resulted in a higher UTS; and

(3) when the fracture occurs through the eutectic region (sample B), the UTS is higher than when the fracture occurs within the IMCs layer (sample D).

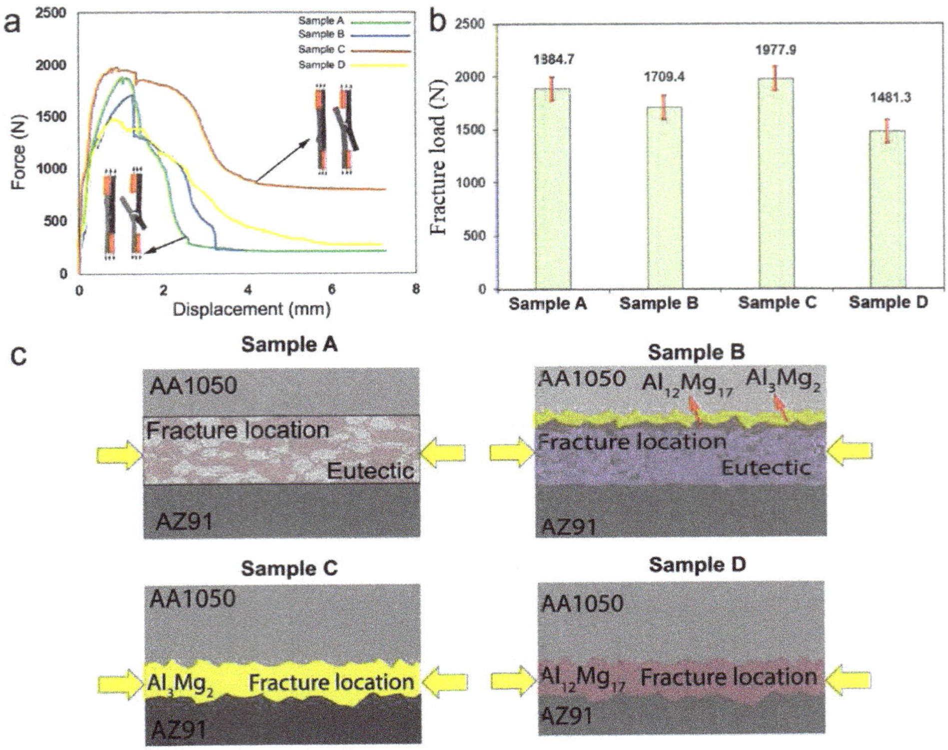

Figure 7. (**a**) Examined shear stress-displacement diagram of the shear-tensile tests of welded samples, (**b**) bar chart demonstrating obtained maximum UTS for samples, and (**c**) a schematic illustration of fracture location in sample A–D.

In order to realize the correlation between microstructure and fracture behavior, the fracture surfaces were investigated by SEM (Figure 8). A comparison of results reveals that fracture location initiation significantly impacts fracture morphology. Intermetallic compounds greatly influence fracture morphology in Al/Mg alloys. Because of their low fracture toughness, the IMCs are prone to cracking and breaking under load [17]. As EDS results proved, $Al_{12}Mg_{17}$ and Al_3Mg_2 are two intermetallic compounds detected at the interface of Al and Mg.

Figure 8(a_1,a_2) show the fracture surfaces of sample A from the AA1050 and AZ91 sides, respectively. The fracture surfaces indicate the presence of both ductile and brittle characteristics, resulting in a lamellar structure. The brittle fracture was observed in the $Al_{12}Mg_{17}$ intermetallic compound in the nugget zone. The failure of sample A occurred from the nugget zone, which includes the Mg phase and eutectic structure comprising the Mg and $Al_{12}Mg_{17}$ intermetallic compound, as shown in Figure 7c. The fracture surfaces of sample B are presented in Figure 8(b_1,b_2). The fracture surface of the AA1050 side shows the presence of both ductile and brittle characteristics, whereas the AZ91 side shows a completely brittle fracture. The brittle fracture occurred between the $Al_{12}Mg_{17}$ and Al_3Mg_2 intermetallic compounds on the Al side, whereas ductile fracture occurred in the AA1050 matrix. The brittle fracture occurred in the $Al_{12}Mg_{17}$ intermetallic compound on the Mg side. The brittle phase in both Al and Mg sides had the composition of $Al_{12}Mg_{17}$ and

Al_3Mg_2. The failure of sample B occurred from the mixing zone, which includes the Mg phase and eutectic structure comprising the Mg and $Al_{12}Mg_{17}$ intermetallic compound, as shown in Figure 7c. The fracture surfaces of sample C are presented in Figure 8(c_1,c_2). The fracture surface of the Al side indicates a brittle fracture, whereas the Mg side shows a ductile fracture. The presence of intermetallic compounds Al_3Mg_2 caused a brittle failure in the welding zone. The failure of sample C occurred from within the mixing zone, which includes the Mg phase and intermetallic compound Al_3Mg_2. According to Figure 5c, the intermetallic compound formed in sample C is in the form of a continuous layer with very low thickness, resulting in the highest strength among the samples. The arrangement of the different phases at the junction is shown in Figure 7c. The fracture surface of sample C is soft, and the strength is very high due to the thinness of the intermetallic compounds, which caused the fracture strength of the weld to exceed the yield strength of the aluminum. Before the weld failure, the base Al experienced some plastic deformation, with a force required to yield Al in this state equal to N 2025 (15 × 3 × 45), resulting in high deformation. The fracture surfaces of sample D are depicted in Figure 8(d_1,d_2). The fracture surface of the AA1050 side indicated the presence of both ductile and brittle characteristics, with the $Al_{12}Mg_{17}$ composition observed on the fracture surface. The recoil of $Al_{12}Mg_{17}$ was higher on the Al side than on the Mg side, resulting in ductile failure in some areas but often in brittle fractures at the joint. On the other hand, the fracture on the AZ91 side was completely brittle and from the grain boundary. The failure of sample D occurred between the Mg and $Al_{12}Mg_{17}$ intermetallic compound, which is continuously formed at the weld boundary, as shown in Figure 5d.

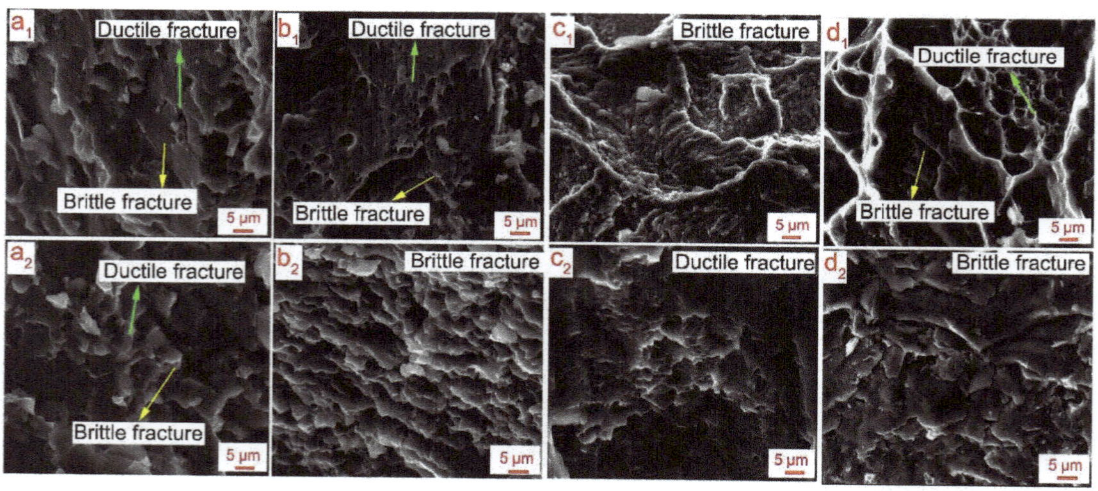

Figure 8. SEM image of (**a_1–d_1**) Al side and (**a_2–d_2**) Mg side of the fracture surface from samples A-D, respectively.

Overall, the presence of intermetallic compounds and the lamellar structure resulted in brittle fracture in some areas, whereas the presence of ductile fracture in other areas is likely due to the plastic deformation of the aluminum matrix. These findings suggest that the microstructure and composition of the materials play a crucial role in determining the fracture behavior of the welded samples.

From the shear-tensile test and fracture morphology, it could be concluded that the relation between brittle fracture and crack propagation is associated with the formed brittle IMCs at the stir interface of AA1050/AZ91, which is consistent with other reports [52–54].

The microhardness distribution across the weld line is shown in Figure 9. The microhardness results show an increase in hardness within the welding zone, followed by a rapid decrease as one moves away from the central weld location. Two distinct hardness distribution profiles were evident on either side.

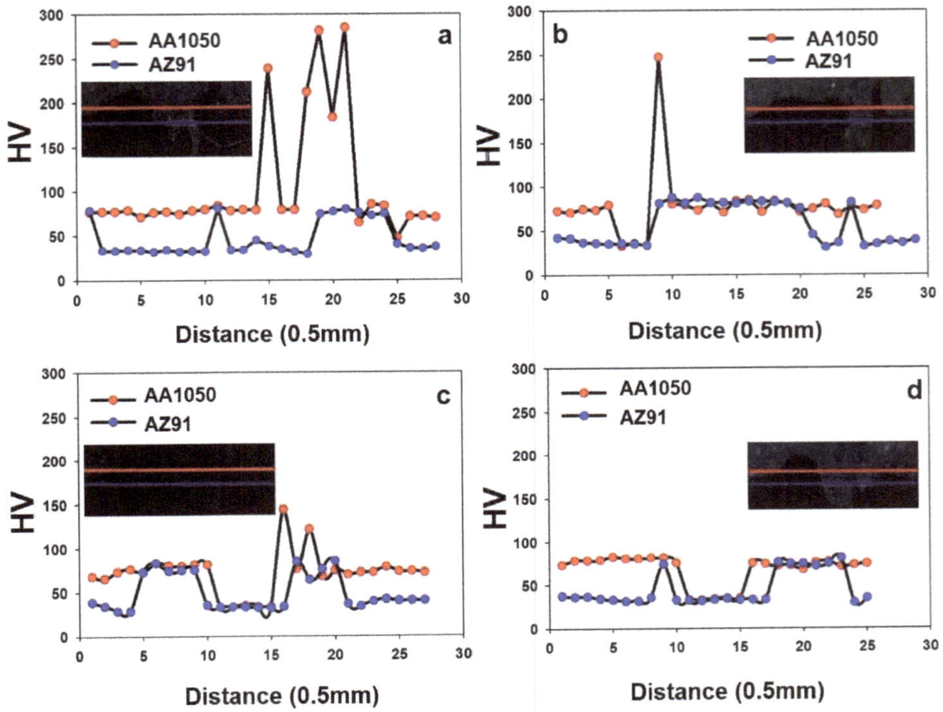

Figure 9. (**a**–**d**) Hardness profile of the joints for sample A–D. The insect picture in panels show the microhardness measurement line of specimens (red line: AA1050 side and blue line: AZ91 side).

Throughout Figure 9, the results indicate that the Al side of sample A has the highest hardness. The presence of intermetallic compounds in the nugget zone caused a sudden increase in hardness in the weld zone, and the accumulation of these compounds resulted in higher hardness compared to the base metal. Furthermore, Figure 9a,b revealed that there was a significant increase in hardness in some locations, which is probably due to the formation of eutectic in the stir zone. The coincidence of hardness values in AA1050 and AZ91 is attributed to drag of the layers into each other. Though SEM analysis revealed the presence of intermetallic compounds such as Al_2Mg_3 and $Al_{12}Mg_{17}$ at the welding boundary in samples C and D, the hardness values are lower in these samples. This is due to two factors. First, the volume of IMCs formed in these samples is lower and second, the IMCs layers are thin and dispersed in the matrix.

Considering the discussed results, the correlation between mechanical properties of the joint and welding parameters can be explained as follows:

At a low rotation rate of the tool, a thin intermetallic compound forms at the Al-Mg interface due to the low heat generation. Therefore, a high level of weld strength can be obtained. In our cases, the fracture surface is almost composed of brittle material.

At a high tool rotation speed, a severe mixing of Al and Mg occurs in the stir zone. The severe mixing along with the high local temperature results in local diffusivity of Al and Mg which in turn promotes liquid $Al_{12}Mg_{17}$-Mg eutectic compound formation. The formation of this liquid compound decreases the friction coefficient between the tool and the material; consequently, the temperature during welding decreases. This decline in the intermetallic growth leads to the formation of a very thin layer of Al-Mg intermetallic compounds at the interface. Whereas this eutectic layer consists of brittle ($Al_{12}Mg_{17}$) and ductile (Mg) layers, the fracture takes place in this region. The EDS study of the fracture mode confirmed the presence of a mixture of ductile and brittle materials.

4. Conclusions

In this study, the influences of FSW welding parameters (welding traverse speed and tool rotation speed) on the weld strength of AA1050 to AZ91 were investigated. To visualize the formation of eutectic and liquefaction structures at the interface of AA1050 and AZ91, SEM and EDS analyses were applied. The main results obtained by friction stir lap welding of AA1050/AZ91 are listed as follows:

- The tool rotation speed plays a major role in the joint quality and intermetallic compounds;
- At low rotation speed of the tool, a thin intermetallic layer forms at the interface, and the joint strength is the highest among all the parameters. The fracture propagates through the AZ91 layer;
- At high rotation speed, liquation takes place in the weld nugget which is composed of a eutectic compound. This liquation causes the slippage of the tool and lowers the heat generation during welding. Moreover, the diffusion rate of elements decreases due to the existence of the eutectic layer. Hence, a thin layer of intermetallic compound forms at the interface. The fracture propagates through the eutectic compound;
- At a higher welding traverse speed with a lower rotational speed, insufficient frictional local heat input and lower exposure time lead to inadequate metallurgical bonding at the AA1050-AZ91 interface. Some defects are observed in this sample due to insufficient material flow;
- The results suggested that rotation rate affects the formation of eutectic structures at the interface;
- The formation of eutectic compounds is influenced by the rotation speed, which is not related to the advanced speed at higher rotation speeds;
- When aluminum penetrates magnesium, it enhances the hardness of the magnesium, whereas magnesium infiltration in aluminum decreases the hardness of the aluminum;
- The formation of eutectic compositions plays a significant role in ensuring consistent hardness levels.

Author Contributions: Conceptualization, R.B.; Investigation, K.T.; Writing—original draft, K.T.; Writing—review & editing, R.B. and G.E.B.; Supervision, L.F.M.d.S. All authors have read and agreed to the published version of the manuscript.

Funding: The authors gratefully thank Arak University for supporting this project. The authors also acknowledge funding under the reference "UIDP/50022/2020—LAETA—Laboratório Associado de Energia, Transportes e Aeronáutica".

Conflicts of Interest: The authors declare no conflict of interest.

References

1. Thirumoorthy, A.; Arjunan, T.; Kumar, K.S. Latest research development in aluminum matrix with particulate reinforcement composites—A review. *Mater. Today Proc.* **2018**, *5*, 1657–1665. [CrossRef]
2. Ng, C.; Yahaya, S.; Majid, A. Reviews on aluminum alloy series and its applications. *J. Acad. Ind. Res.* **2017**, *5*, 708–716.
3. Baghdadi, A.H.; Rajabi, A.; Selamat, N.F.M.; Sajuri, Z.; Omar, M.Z. Effect of post-weld heat treatment on the mechanical behavior and dislocation density of friction stir welded Al6061. *Mater. Sci. Eng. A* **2019**, *754*, 728–734. [CrossRef]
4. He, L.; Pan, L.; Zhou, W.; Niu, Z.; Chen, X.; Chen, M.; Zhang, Q.; Pan, W.; Xiao, P.; Li, Y. Thermal corrosion behavior of $Yb_4Hf_3O_{12}$ ceramics exposed to calcium-ferrum-alumina-silicate (CFAS) at 1400 °C. *J. Eur. Ceram. Soc.* **2023**, *43*, 4114–4123. [CrossRef]
5. Pan, L.; He, L.; Niu, Z.; Xiao, P.; Zhou, W.; Li, Y. Corrosion behavior of ytterbium hafnate exposed to water-vapor with $Al(OH)_3$ impurities. *J. Eur. Ceram. Soc.* **2023**, *43*, 612–620. [CrossRef]

6. Beygi, R.; Talkhabi, A.A.; Mehrizi, M.Z.; Marques, E.A.; Carbas, R.J.; da Silva, L.F. A Novel Lap-Butt Joint Design for FSW of Aluminum to Steel in Tee-Configuration: Joining Mechanism, Intermetallic Formation, and Fracture Behavior. *Metals* **2023**, *13*, 1027. [CrossRef]
7. Singh, V.P.; Patel, S.K.; Ranjan, A.; Kuriachen, B. Recent research progress in solid state friction-stir welding of aluminium–magnesium alloys: A critical review. *J. Mater. Res. Technol.* **2020**, *9*, 6217–6256. [CrossRef]
8. Zhou, C.; Yang, X.; Luan, G. Effect of root flaws on the fatigue property of friction stir welds in 2024-T3 aluminum alloys. *Mater. Sci. Eng. A* **2006**, *418*, 155–160. [CrossRef]
9. Laska, A.; Szkodo, M.; Cavaliere, P.; Perrone, A. Influence of the Tool Rotational Speed on Physical and Chemical Properties of Dissimilar Friction-Stir-Welded AA5083/AA6060 Joints. *Metals* **2022**, *12*, 1658. [CrossRef]
10. Shelley, T. Joining forces for multiple properties. *Eureka* **2003**.
11. Habibnia, M.; Shakeri, M.; Nourouzi, S.; Givi, M.B. Microstructural and mechanical properties of friction stir welded 5050 Al alloy and 304 stainless steel plates. *Int. J. Adv. Manuf. Technol.* **2015**, *76*, 819–829. [CrossRef]
12. Liu, Q.; Han, R.; Gao, Y.; Ke, L. Numerical investigation on thermo-mechanical and material flow characteristics in friction stir welding for aluminum profile joint. *Int. J. Adv. Manuf. Technol.* **2021**, *114*, 2457–2469. [CrossRef]
13. Gaurav, S.; Mishra, R.; Zunaid, M. A critical review on mechanical and microstructural properties of dissimilar aluminum (Al)-magnesium (Mg) alloys. *J. Adhes. Sci. Technol.* **2022**, 1–33. [CrossRef]
14. Fu, B.; Qin, G.; Li, F.; Meng, X.; Zhang, J.; Wu, C. Friction stir welding process of dissimilar metals of 6061-T6 aluminum alloy to AZ31B magnesium alloy. *J. Mater. Process. Technol.* **2015**, *218*, 38–47. [CrossRef]
15. Xu, Y.; Ke, L.; Mao, Y.; Yang, P.; Niu, P. Friction stir welding of aluminium to magnesium: A critical review. *Mater. Sci. Technol.* **2022**, *38*, 517–534. [CrossRef]
16. Elangovan, K.; Balasubramanian, V.; Valliappan, M. Effect of tool pin profile and tool rotational speed on mechanical properties of friction stir welded AA6061 aluminium alloy. *Mater. Manuf. Process.* **2008**, *23*, 251–260. [CrossRef]
17. Nonnenmann, T.; Beygi, R.; Carbas, R.J.; da Silva, L.F.; Öchsner, A. Feasibility study on hybrid weld-bonding between dissimilar material for automotive industry. *Int. J. Adhes. Adhes.* **2023**, *121*, 103316. [CrossRef]
18. Safeen, M.W.; Russo Spena, P. Main issues in quality of friction stir welding joints of aluminum alloy and steel sheets. *Metals* **2019**, *9*, 610. [CrossRef]
19. El-Sayed, M.M.; Shash, A.; Abd-Rabou, M.; ElSherbiny, M.G. Welding and processing of metallic materials by using friction stir technique: A review. *J. Adv. Join. Process.* **2021**, *3*, 100059. [CrossRef]
20. Meschut, G.; Merklein, M.; Brosius, A.; Drummer, D.; Fratini, L.; Füssel, U.; Gude, M.; Homberg, W.; Martins, P.; Bobbert, M. Review on mechanical joining by plastic deformation. *J. Adv. Join. Process.* **2022**, *5*, 100113. [CrossRef]
21. Mohammadi, J.; Behnamian, Y.; Mostafaei, A.; Izadi, H.; Saeid, T.; Kokabi, A.; Gerlich, A. Friction stir welding joint of dissimilar materials between AZ31B magnesium and 6061 aluminum alloys: Microstructure studies and mechanical characterizations. *Mater. Charact.* **2015**, *101*, 189–207. [CrossRef]
22. Shi, L.; Wu, C. Transient model of heat transfer and material flow at different stages of friction stir welding process. *J. Manuf. Process.* **2017**, *25*, 323–339. [CrossRef]
23. Husain, M.M.; Sarkar, R.; Pal, T.; Prabhu, N.; Ghosh, M. Friction stir welding of steel: Heat input, microstructure, and mechanical property co-relation. *J. Mater. Eng. Perform.* **2015**, *24*, 3673–3683. [CrossRef]
24. Krishnan, M.; Subramaniam, S.K. Investigation of mechanical and metallurgical properties of friction stir corner welded dissimilar thickness AA5086-AA6061 aluminium alloys. *Mater. Res.* **2018**, *21*. [CrossRef]
25. Liu, F.; Fu, L.; Chen, H. Effect of high rotational speed on temperature distribution, microstructure evolution, and mechanical properties of friction stir welded 6061-T6 thin plate joints. *Int. J. Adv. Manuf. Technol.* **2018**, *96*, 1823–1833. [CrossRef]
26. Salih, O.S.; Ou, H.; Wei, X.; Sun, W. Microstructure and mechanical properties of friction stir welded AA6092/SiC metal matrix composite. *Mater. Sci. Eng. A* **2019**, *742*, 78–88. [CrossRef]
27. Buffa, G.; Baffari, D.; Di Caro, A.; Fratini, L. Friction stir welding of dissimilar aluminium–magnesium joints: Sheet mutual position effects. *Sci. Technol. Weld. Join.* **2015**, *20*, 271–279. [CrossRef]
28. Liu, H.; Fujii, H.; Maeda, M.; Nogi, K. Mechanical properties of friction stir welded joints of 1050–H24 aluminium alloy. *Sci. Technol. Weld. Join.* **2003**, *8*, 450–454. [CrossRef]
29. Beygi, R.; Akhavan-Safar, A.; Carbas, R.; Barbosa, A.; Marques, E.; da Silva, L. Utilizing a ductile damage criterion for fracture analysis of a dissimilar aluminum/steel joint made by friction stir welding. *Eng. Fract. Mech.* **2022**, *274*, 108775. [CrossRef]
30. Zhao, Y.; Jiang, S.; Yang, S.; Lu, Z.; Yan, K. Influence of cooling conditions on joint properties and microstructures of aluminum and magnesium dissimilar alloys by friction stir welding. *Int. J. Adv. Manuf. Technol.* **2016**, *83*, 673–679. [CrossRef]
31. Lv, X.; Wu, C.; Padhy, G. Diminishing intermetallic compound layer in ultrasonic vibration enhanced friction stir welding of aluminum alloy to magnesium alloy. *Mater. Lett.* **2017**, *203*, 81–84. [CrossRef]
32. Kumar, N.; Yuan, W.; Mishra, R. Challenges and opportunities for friction stir welding of dissimilar alloys and materials (Chapter 7). In *Friction Stir Welding of Dissimilar Alloys and Materials*; Elsevier Butterworth-Heinemann: Oxford, UK, 2015; pp. 123–126.
33. Paul, A.R.; Mukherjee, M.; Singh, D. A critical review on the properties of intermetallic compounds and their application in the modern manufacturing. *Cryst. Res. Technol.* **2022**, *57*, 2100159. [CrossRef]
34. Shah, L.; Othman, N.; Gerlich, A. Review of research progress on aluminium–magnesium dissimilar friction stir welding. *Sci. Technol. Weld. Join.* **2018**, *23*, 256–270. [CrossRef]

35. Yang, Y.; Dong, H.; Kou, S. Liquation tendency and liquid-film formation in friction stir spot welding. *Weld. J.* **2008**, *87*, 202S–211S.
36. Yamamoto, N.; Liao, J.; Watanabe, S.; Nakata, K. Effect of intermetallic compound layer on tensile strength of dissimilar friction-stir weld of a high strength Mg alloy and Al alloy. *Mater. Trans.* **2009**, *50*, 2833–2838. [CrossRef]
37. Firouzdor, V.; Kou, S. Al-to-Mg friction stir welding: Effect of positions of Al and Mg with respect to the welding tool. *Weld. J.* **2009**, *88*, 213–224.
38. Prabha, K.A.; Putha, P.K.; Prasad, B.S. Effect of tool rotational speed on mechanical properties of aluminium alloy 5083 weldments in friction stir welding. *Mater. Today Proc.* **2018**, *5*, 18535–18543. [CrossRef]
39. Ji, S.; Meng, X.; Liu, Z.; Huang, R.; Li, Z. Dissimilar friction stir welding of 6061 aluminum alloy and AZ31 magnesium alloy assisted with ultrasonic. *Mater. Lett.* **2017**, *201*, 173–176. [CrossRef]
40. Rao, H.; Yuan, W.; Badarinarayan, H. Effect of process parameters on mechanical properties of friction stir spot welded magnesium to aluminum alloys. *Mater. Des.* **2015**, *66*, 235–245. [CrossRef]
41. Firouzdor, V.; Kou, S. Al-to-Mg friction stir welding: Effect of material position, travel speed, and rotation speed. *Metall. Mater. Trans. A* **2010**, *41*, 2914–2935. [CrossRef]
42. Beygi, R.; Zarezadeh Mehrizi, M.; Akhavan-Safar, A.; Mohammadi, S.; da Silva, L.F. A Parametric Study on the Effect of FSW Parameters and the Tool Geometry on the Tensile Strength of AA2024–AA7075 Joints: Microstructure and Fracture. *Lubricants* **2023**, *11*, 59. [CrossRef]
43. Kumar, U.; Acharya, U.; Saha, S.C.; Saha Roy, B. Microstructure and mechanical property of friction stir welded Al-Mg joints by adopting modified joint configuration technique. *Mater. Today Proc.* **2020**, *26*, 2083–2088. [CrossRef]
44. Wu, H.; Chen, Y.-C.; Strong, D.; Prangnell, P. Stationary shoulder FSW for joining high strength aluminum alloys. *J. Mater. Process. Technol.* **2015**, *221*, 187–196. [CrossRef]
45. Beygi, R.; Pouraliakbar, H.; Torabi, K.; Fallah, V.; Kim, S.; Shi, R.; da Silva, L. The inhibitory effect of stir zone liquefaction and eutectic-phase formation on the growth of γ/β intermetallics during dissimilar FSW of Al/Mg alloys. *J. Manuf. Process.* **2021**, *70*, 152–162. [CrossRef]
46. Ji, S.; Li, Z.; Zhang, L.; Zhou, Z.; Chai, P. Effect of lap configuration on magnesium to aluminum friction stir lap welding assisted by external stationary shoulder. *Mater. Des.* **2016**, *103*, 160–170. [CrossRef]
47. Ghiasvand, A.; Ranjbarnodeh, E.; Mirsalehi, S.E. The microstructure and mechanical properties of single-pass and double-pass lap joint of Al 5754H-11 and Mg AZ31-O alloys by friction stir welding. *J. Mater. Res. Technol.* **2023**, *23*, 6023–6038. [CrossRef]
48. Akbari, M.; Aliha, M.; Keshavarz, S.; Bonyadi, A. Effect of tool parameters on mechanical properties, temperature, and force generation during FSW. *Proc. Inst. Mech. Eng. Part L J. Mater. Des. Appl.* **2019**, *233*, 1033–1043. [CrossRef]
49. Mao, Y.; Yang, P.; Zhang, W.; Li, N.; Nie, H.; Lin, D.; Ke, L. Improving tensile-shear properties of friction stir lap welded dissimilar Al/Mg joints by eliminating hook defect and controlling interfacial reaction. *Chin. J. Aeronaut.* **2022**, in press. [CrossRef]
50. Lakshminarayanan, A.; Annamalai, V. Fabrication and performance evaluation of dissimilar magnesium–aluminium alloy multi-seam friction stir clad joints. *Trans. Nonferrous Met. Soc. China* **2017**, *27*, 25–35. [CrossRef]
51. Chen, W.; Wang, W.; Liu, Z.; Zhai, X.; Bian, G.; Zhang, T.; Dong, P. Improvement in tensile strength of Mg/Al alloy dissimilar friction stir welding joints by reducing intermetallic compounds. *J. Alloys Compd.* **2021**, *861*, 157942. [CrossRef]
52. Sato, Y.S.; Park, S.H.C.; Michiuchi, M.; Kokawa, H. Constitutional liquation during dissimilar friction stir welding of Al and Mg alloys. *Scr. Mater.* **2004**, *50*, 1233–1236. [CrossRef]
53. Jagadeesha, C. Dissimilar friction stir welding between aluminum alloy and magnesium alloy at a low rotational speed. *Mater. Sci. Eng. A-Struct. Mater. Prop. Microstruct. Process.* **2014**, *616*, 55–62. [CrossRef]
54. Shi, H.; Chen, K.; Liang, Z.; Dong, F.; Yu, T.; Dong, X.; Zhang, L.; Shan, A. Intermetallic compounds in the banded structure and their effect on mechanical properties of Al/Mg dissimilar friction stir welding joints. *J. Mater. Sci. Technol.* **2017**, *33*, 359–366. [CrossRef]

applied sciences

MDPI

Article

Evolutionary Algorithm to Optimize Process Parameters of Al/Steel Magnetic Pulse Welding

Jiyeon Shim [1] and Illsoo Kim [2,*]

1 Carbon & Light Materials Application R&D Group, Korea Institute of Industrial Technology, Jeonju-si 54853, Republic of Korea; shimjy@kitech.re.kr

2 Department of Mechanical Engineering, Mokpo National University, 1666 Youngsan-ro, Chungkye-myun, Muan-gun 58554, Republic of Korea

* Correspondence: ilsookim@mokpo.ac.kr

Abstract: The Magnetic Pulse Welding (MPW) process uses only electromagnetic force to create a solid-state metallurgical bond between a working coil and outer workpiece. The electromagnetic force drives the outer tube to collide with the inner rod, resulting in successful bonding. However, due to the dissimilarity of the MPW joint, only a portion of the interface forms a metallurgical bond, which affects the quality of the joint. Therefore, the purpose of this study is to analyze the effects of process parameters on joint quality through experimental work using RSM. Furthermore, an optimization algorithm is utilized to optimize the process parameters used in magnetic pulse welding. A1070 aluminum and S45C carbon steel were used as the materials, while peak current, gap between working coil and outer tube, and frequency were chosen as the process parameters for MPW. The welding conditions are determined through experimental design. After welding, the maximum load and weld length are measured to analyze the effect of the process parameters, and a prediction model is developed. Specifically, to achieve a high-quality joint, the process parameters are optimized using the Imperialist Competitive Algorithm (ICA) and Genetic Algorithm (GA). The results reveal that the peak current is a significant parameter, and the developed prediction model exhibits high accuracy. Furthermore, the ICA algorithm proves very effective in determining the process parameters for achieving a high-quality Al/Steel MPW joint.

Keywords: magnetic pulse welding; evolutionary algorithm; electromagnetic force; process parameters; Imperialist Competitive Algorithm; Genetic Algorithm

Citation: Shim, J.; Kim, I. Evolutionary Algorithm to Optimize Process Parameters of Al/Steel Magnetic Pulse Welding. *Appl. Sci.* **2023**, *13*, 12881. https://doi.org/10.3390/app132312881

Academic Editors: Ivan Galvão, Rui Manuel Leal and Gustavo H. S. F. L. Carvalho

Received: 31 October 2023
Revised: 20 November 2023
Accepted: 25 November 2023
Published: 30 November 2023

1. Introduction

In the field of automotive engineering, a main challenge facing automakers is to reduce harmful greenhouse gas emissions by decreasing fuel consumption [1,2]. One effective solution for this is to decrease the weight of automotive body assemblies. However, it is crucial to ensure that any weight reduction in the body assembly does not compromise passenger safety. Therefore, lightweight materials with high strength, such as aluminum and magnesium, have been implemented in automotive body construction. Nonetheless, the use of only lightweight materials may mean that specific components do not meet required specifications, so there is a need to develop integration processes for dissimilar materials.

Certain welding processes have been considered for integrating dissimilar materials. Fusion welding methods, such as conventional welding, laser beam welding, and resistance welding, have been studied. Zhao et al. [3] and Chen et al. [4] aimed to enhance the synergy between strength and ductility while also improving the corrosion resistance of a dissimilar joint between Inconel 718 and 316L. Pan et al. [5] conducted a study on the formation mechanisms of intermetallic compounds in a dissimilar joint between aluminum and steel. This investigation was carried out using resistance spot welding. However, the high heat input in these fusion welding processes caused not only thick intermetallic compound (IMC) layers between materials but also cracking, porosity, and oxidation [5]. For this reason, friction stir welding, explosive welding, and magnetic pulse welding have recently been widely studied as methods of solid-state welding. Yang et al. [6] conducted a numerical analysis to investigate the process of mass transfer and material mixing in friction stir welding of dissimilar joints. Bhattacharya et al. [7] attempted to join aluminum alloy to DHP copper using friction stir welding. This process has the capability to produce high-quality joints between different combinations of materials. However, due to tool conditions, several types of defects may occur in joints, such as thickness reduction and flash and keyhole defects [8]. Explosive welding utilizes explosive force to create a successful metallurgical bond between workpieces. Kumar et al. [9] reported on the welding of magnesium and aluminum alloys using an inclined arrangement in explosive welding, while Carvalho et al. [10] studied the effect of flyer material on interface phenomena in explosive welds of aluminum and copper. However, controlling the process parameters is difficult, and experimental work can be unsafe. For this reason, Magnetic Pulse Welding (MPW), a solid-state welding process, has been used in the development of high-quality dissimilar joints. MPW is a high-velocity welding process using an electromagnetic force to cause high-velocity collisions between dissimilar materials, resulting in true metallurgical bonds. MPW uses only electromagnetic force, without filler wire, as shown in Figure 1. It is an environmentally friendly welding process [11,12]. When using MPW for the welding of dissimilar materials such as aluminum and steel, a challenge for researchers is selecting the proper welding parameters to create strong joints with excellent properties. The main process parameters of MPW, including input current, gap between workpieces, and frequency, play important roles in determining joint quality. One weakness of MPW is its potentially short metallurgical bond length. Thus, analysis of the effect of process parameters on weld length is important to obtain high-quality joints.

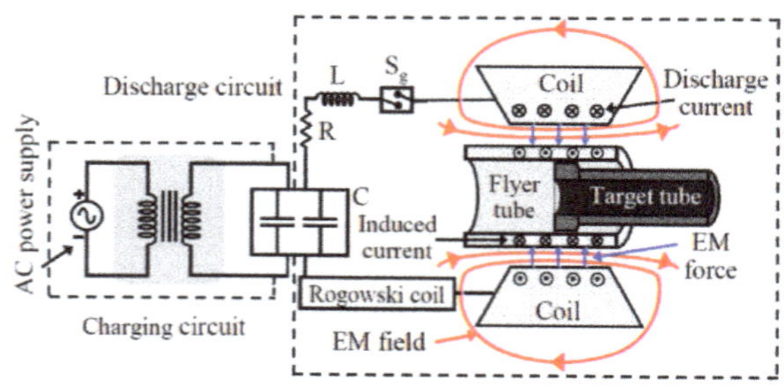

Figure 1. Schematic layout of magnetic pulse welding [12].

Zhang et al. [13] characterized and analyzed the interface structure of an Al/Ta joint, while Geng et al. [14] analyzed the fatigue fracture properties of Al/steel joints using SEM, TEM, and EDS. Recently, Patra et al. [15] performed non-destructive evaluation and investigated the corrosion of Al/steel joints through compositional analysis. Yao et al. [16] discussed the mechanical properties and element diffusion in an additively manufactured 316L to AA5052 MPW joint. This process utilized electromagnetic force, making numerical analysis highly effective for understanding and applying it on site. As a result, researchers have conducted studies using numerical simulations. Shim et al. [17] used a high-accuracy FE model to investigate the characteristics of Al/steel magnetic pulse tubular joints according to discharge time. Bembalge et al. [18] analyzed the effects of gaps between workpieces and of input voltage energy on the mechanical and microstructural properties of joints. Yan et al. [19], utilizing numerical simulations, performed magnetic pulse welding (MPW) with a multiple-seam field-shaper to generate Al/steel joints with uniform circumferential deformation distribution. However, optimizing the MPW process using experimental and numerical methods is challenging due to the need for a large number of trials.

Nowadays, optimization methods are widely used to address problems in welding processes. Notably, response surface methodology (RSM) and the Taguchi method [20,21] are commonly used to determine the relationship between input parameters and output responses, such as joint quality. These methods are effective in minimizing the number of experiments needed for analysis. Shim et al. [1] used RSM to develop a prediction model for burst pressure in dissimilar joints, such as Al/steel and Al/Cu. Ayaz et al. [22] used RSM to investigate the effects of process parameters on the microstructural evolution and mechanical properties of stainless steel and copper joints.

Several evolutionary algorithms have been developed to address optimization problems across a range of scientific and engineering disciplines. These algorithms include the Genetic Algorithm (GA) [23–27], the ant colony optimization algorithm, the bee algorithm, and the Imperialist Competitive Algorithm (ICA) [28–32]. These algorithms establish robust process parameter combinations for achieving superior joint quality. However, the application of evolutionary algorithms to select optimal process parameters in the MPW process has not been explored. Therefore, the objective of this study is to use an evolutionary algorithm to determine the optimal process parameters for achieving maximum weld length. For this, the peak current, gap between workpieces, and frequency were selected as input parameters, while the maximum load and weld length were chosen as output responses. The relationship between the input parameters and the output responses was first examined. Then, a prediction model was developed for application to the evolutionary algorithm. Both ICA and GA were employed as evolutionary algorithms; the determined optimal process parameters were confirmed through verification experiments.

2. Materials and Methods

2.1. Design of the Experiment

The process parameters, such as peak current, gap between workpieces, and frequency, were selected to investigate their effects on the maximum load and weld length. An easy way to estimate the RSM, central composite design (CCD), was employed, and a total of 20 welding conditions were designed to develop the prediction model. The total number of experiments for this study was 20, with three factors (n = 3) and five coded levels. The calculation for this number is based on the expression: (=8: factor points) + 2n (23 = 6 axial points) + 6 (center points: six replications) [1]. Table 1 lists the design parameters and their corresponding five levels. At the low level (−1), peak current, gap between workpiece, and frequency were set to 350 kA, 0.7 mm, and 17 kHz; at the high level (+1), these were set to 450 kA, 1.0 mm, and 20 kHz, respectively, as shown in Table 2.

Table 1. Process parameters and their levels.

Parameter	Unit	Symbol	Level				
			−2	−1	0	1	+2
Peak current	kA	C	316	350	400	450	484
Gap between workpieces	mm	G	0.2	0.4	0.7	1.0	1.2
Frequency	kHz	F	16.0	17	18.5	20	21.0

Table 2. Design of experiment conditions based on CCD.

Run	Input Parameters			Run	Input Parameters		
	C	G	F		C	G	F
1	400	0.7	18.5	11	400	0.7	18.5
2	400	0.7	18.5	12	450	1.0	20.0
3	400	0.2	18.5	13	350	1.0	20.0
4	450	0.4	17.0	14	400	0.7	18.5

Table 2. *Cont.*

Run	Input Parameters			Run	Input Parameters		
	C	G	F		C	G	F
5	316	0.7	18.5	15	400	0.7	16.0
6	484	0.7	18.5	16	350	0.4	20.0
7	350	0.4	17.0	17	400	0.7	18.5
8	400	1.2	18.5	18	400	0.7	18.5
9	450	1.0	17.0	19	450	0.4	20.0
10	350	1.0	17.0	20	400	0.7	21.0

2.2. Experimental Procedure

This study utilized the magnetic pulse welding system produced by WELMATE Co., Ltd., Cheonan-si, Republic of Korea as shown in Figure 2. This system consists of a pulse power source, a working coil, and a jig/fixture part for welding.

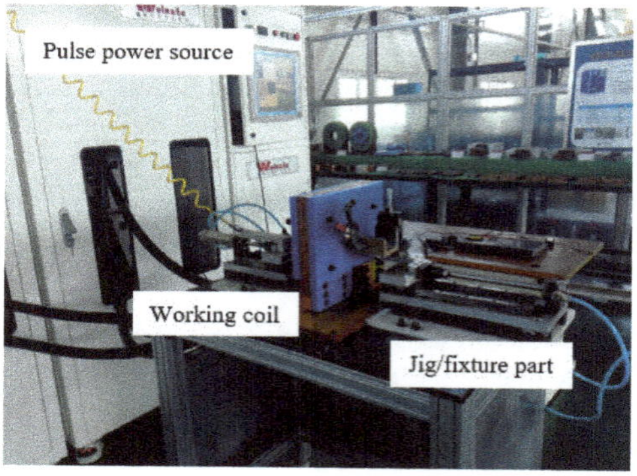

Figure 2. Magnetic pulse welding system.

The pulse power source in this study employed capacitors. Specifically, eight capacitors were connected in parallel, allowing for individual operation. This setup enabled variations in peak current and frequency. Eight triggered switch units were utilized to discharge the capacitors into the coil. The total capacitance of the system was 512 μF, and the charging voltage reached 15 kV. Additionally, to measure the waveform, a Rogowski coil was placed at the junction where the pulse power source and the working coil connect.

The workpieces themselves included an outer workpiece made of an A1070 aluminum tube with a thickness of 0.8 mm. The inner workpiece was made of S45C steel rod. Gaps between workpieces varied depending on the experimental conditions, as detailed in Table 2. The chemical compositions of the workpieces can be found in Table 3.

Table 3. Chemical composition of workpieces (wt.%).

Material	Si	Ga	Ti	Fe	Zn	Al	P	Mn	Cr	C	Fe
A1070	0.07	0.01	0.01	0.20	0.01	Bal	-	-	-	-	-
S45C	0.23	-	-	-	-	-	0.02	0.02	0.08	0.48	Bal

After conducting magnetic pulse welding under various welding conditions, tensile testing and optical microscopy were used to measure the maximum load and weld length, as shown in Figure 3.

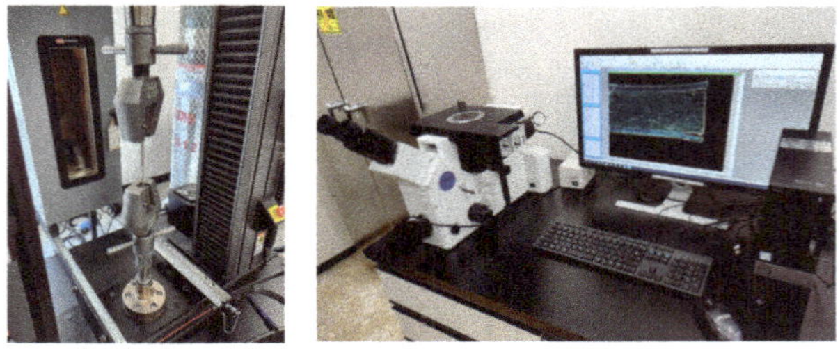

(a) (b)

Figure 3. Setup for joint quality evaluation test. (**a**) Measurement of the maximum load. (**b**) Observation of weld length.

Table 4 presents waveforms obtained under different welding conditions. This process employed only a quarter of the cycle, precisely until the peak current of the measured waveform [17]. System control allowed for variation of both peak current and frequency of the damped sinusoidal waveform.

Table 4. Discharge waveforms (X: 20 µs/div, Y: 250 kA/div).

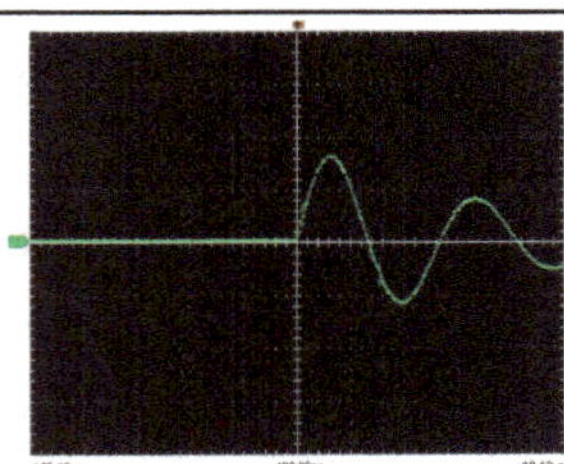	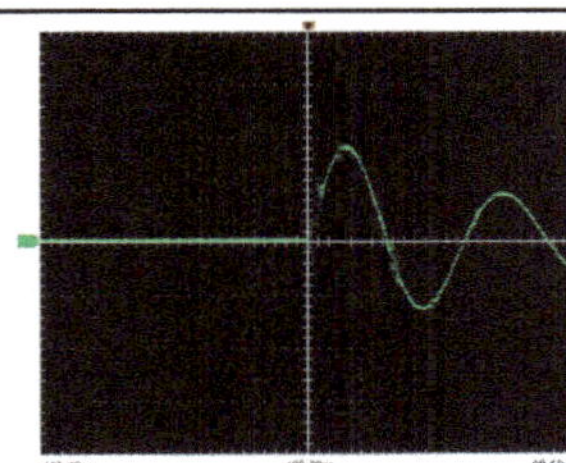	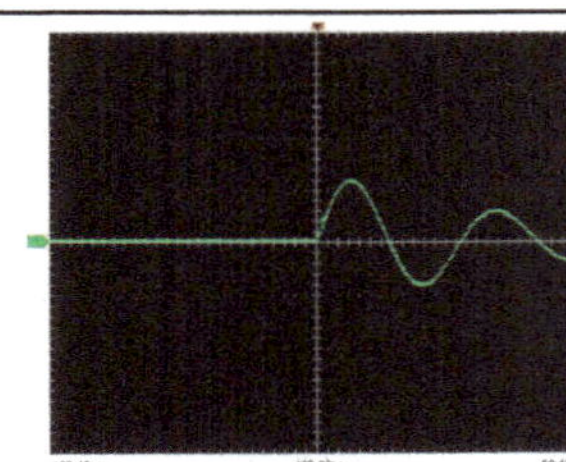
400 kA, 18.5 kHz	450 kA 17 kHz	316 kA 18.5 kHz
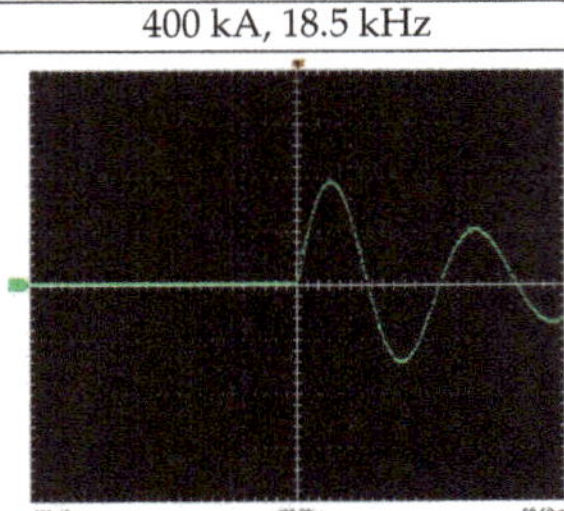	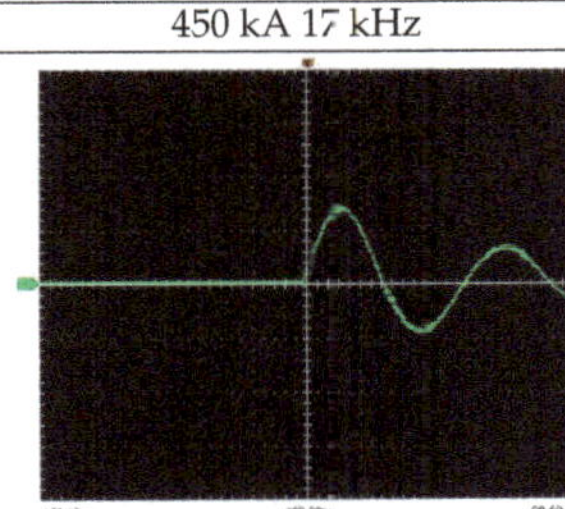	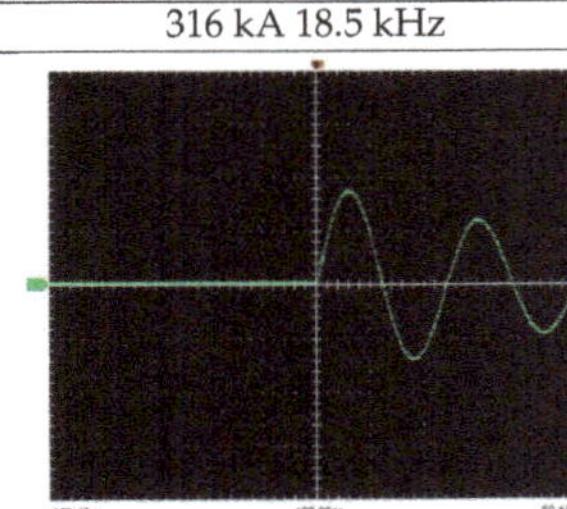
484 kA 18.5 kHz	370 kA 17 kHz	450 kA 20 kHz
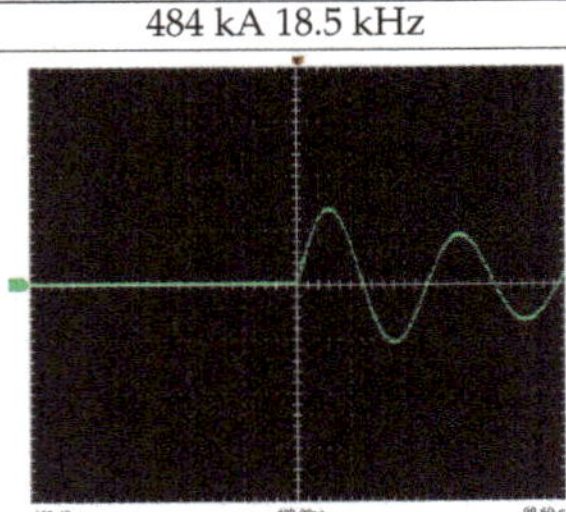	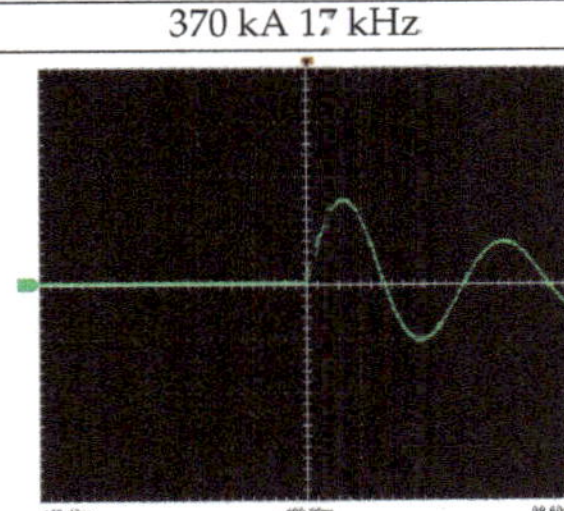	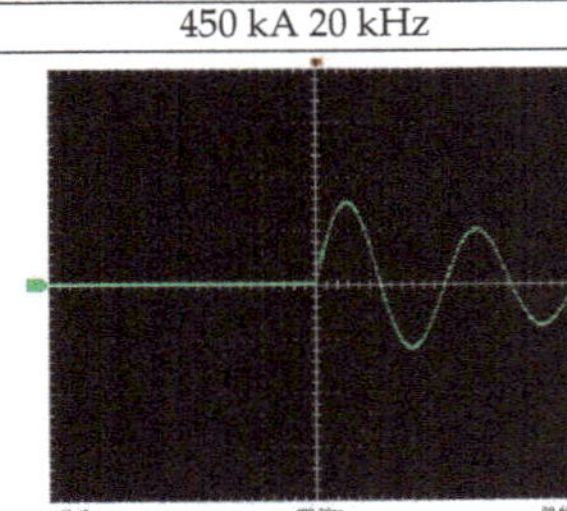
350 kA 20 kHz	450 kA 15 kHz	400 kA 21 kHz

2.3. Results and Discussion

Results of tensile testing and observation of weld length in Tables 5 and 6 show cross sections of joints and maximum loads, respectively.

Table 5. Cross sections of Al/Steel joints.

No.	Cross Section	No.	Cross Section	No.	Cross Section	No.	Cross Section
1	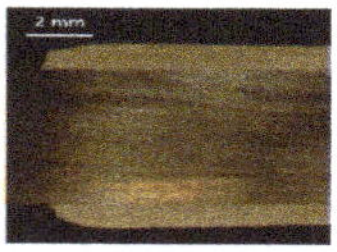	6	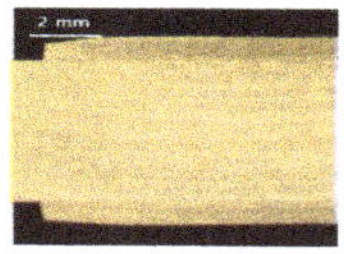	11	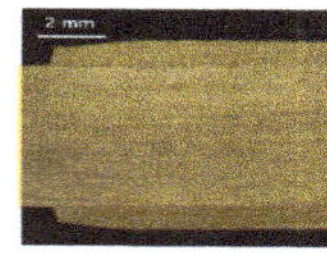	16	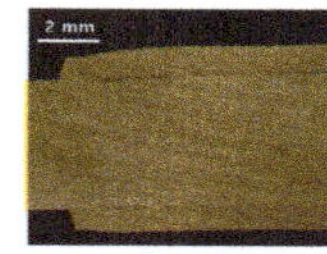
2		7	Separation	12	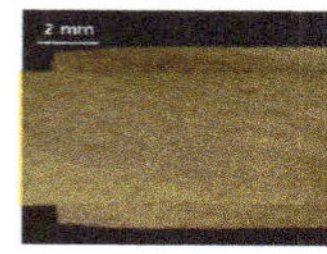	17	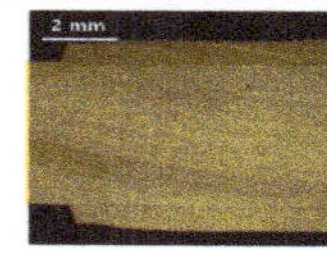
3	Separation	8	Separation	13	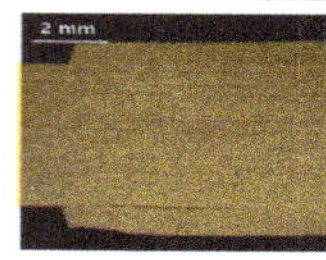	18	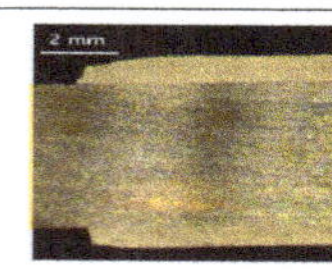
4	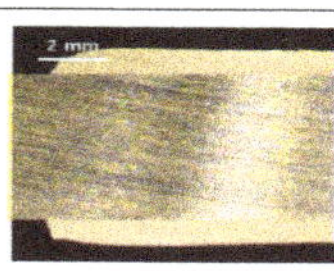	9	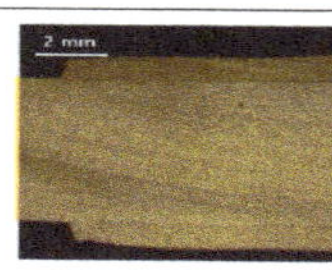	14	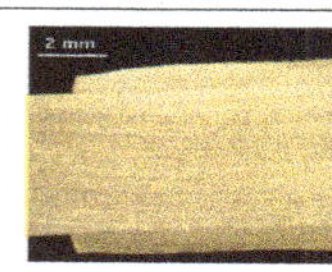	19	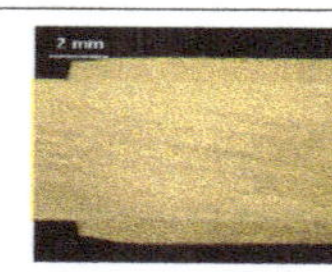
5	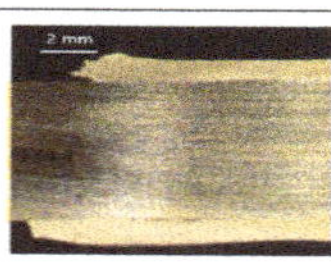	10	Separation	15		20	

Table 6. Experimental results.

Trial No.	Input Parameters			Responses	
	C	G	F	Max. Load	Weld Length
1	400	0.7	18.5	1.079	0.27
2	400	0.7	18.5	1.140	0.23
3	400	0.2	18.5	0.043	-
4	450	0.4	17.0	0.785	0.20
5	316	0.7	18.5	0.260	-
6	484	0.7	18.5	1.825	0.54
7	350	0.4	17.0	0.120	-
8	400	1.2	18.5	0.192	-
9	450	1.0	17.0	1.055	0.17
10	350	1.0	17.0	0.042	-
11	400	0.7	18.5	1.143	0.29
12	450	1.0	20.0	1.309	0.50
13	350	1.0	20.0	0.501	0.09
14	400	0.7	18.5	1.069	0.25
15	400	0.7	16.0	0.657	0.10
16	350	0.4	20.0	0.428	0.02
17	400	0.7	18.5	1.067	0.25
18	400	0.7	18.5	1.080	0.27
19	450	0.4	20.0	1.106	0.41
20	400	0.7	21.0	1.467	0.41

2.3.1. Effects of Process Parameters

An analysis of variance (ANOVA) was used to analyze the impacts of parameters on the maximum load and weld length of the joint. Table 7 presents ANOVA results for the mean values of maximum load. The table includes the following symbols: (DF) for degrees of freedom, (Adj SS) for adjusted sum of squares, (Seq SS) for the sequential sum of squares, (Adj MS) for adjusted mean squares, (R-Sq) for coefficient of determination, 'F' for the variance or Fisher ratio of any factor, (Seq SS) for the pure sum of squares of any factor, and 'P' for critical probability. A *p*-value below 0.05 indicates that the model is statistically significant [33]. The results indicate that parameter C has a significant effect on the maximum load, with a contribution of 48.7%. Parameter F follows with a contribution of 10.6%, and parameter G has a contribution of 0.5%. On the other hand, interaction terms such as C × G, G × G, and G × F have negligible effects on the maximum load. The coefficient of determination (R^2) for the developed prediction models is 96.02%. MPW is a high-velocity collision welding process.

Table 7. ANOVA table for maximum load.

Source	DF	Adj SS	Adj MS	F-Value	*p*-Value
Model	9	4.98741	0.55416	101.40	0.000
Linear	3	3.02280	1.00760	184.37	0.000
C	1	2.45984	2.45984	450.09	0.000
G	1	0.02748	0.02748	5.03	0.049
F	1	0.53548	0.53548	97.98	0.000
Square	3	1.93055	0.64352	117.75	0.000
Interaction	3	0.03405	0.01135	2.08	0.167
Error	10	0.05465	0.00547		
Total	19	5.04206			

Based on the ANOVA results, a quadratic polynomial model for maximum load was developed using coded units (i.e., −1, 0, +1). The model can be represented by the following equation:

$$\begin{aligned}\text{Maximumload} = & -10.78 - 0.0152C + 3.08G - 0.521F - 0.000015C^2 - 3.868G^2 - 0.0135F^2 \\ & +0.00342CG + 0.000160CF + 0.0600GF\end{aligned} \quad (1)$$

Table 8 presents the ANOVA results for weld length. The results indicate that parameters C and F have significant effects on weld length. Specifically, parameter C has the greatest contribution at 55.1%, followed by parameter F at 17%, and parameter G at 0.2%. Additionally, the interaction terms C × F and C × G influence weld length. The contribution of C × F is 4%, while C × G contributes 0.7%. The developed prediction models have a coefficient of determination (R^2) of 95.81%.

Table 8. ANOVA table for weld length.

Source	DF	Adj SS	Adj MS	F-Value	*p*-Value
Model	9	0.568492	0.063166	140.12	0.000
Linear	3	0.417941	0.139314	309.04	0.000
C	1	0.316236	0.316236	701.52	0.000
G	1	0.001237	0.001237	2.75	0.129
F	1	0.100468	0.100468	222.87	0.000
Square	3	0.122914	0.040971	90.89	0.000
Interaction	3	0.027637	0.009212	20.44	0.000
C × G	1	0.000012	0.000012	0.03	0.871
C × F	1	0.023112	0.023112	51.27	0.000
G × F	1	0.004512	0.004512	10.01	0.010
Error	10	0.004508	0.000451		
Total	19	0.573000			

The ANOVA results led to a quadratic polynomial model for weld length using coded units (−1, 0, +1). The model is represented by the following equation:

$$\begin{aligned}\text{Weldlength} = 2.07 - 0.00568\text{C} + 1.005\text{G} - 0.244\text{F} - 0.000002C^2 - 1.008G^2 + 0.00133F^2 \\ -0.00058CG + 0.000583CF + 0.0361GF\end{aligned} \quad (2)$$

The peak current and frequency are closely related to the force required for collision with workpieces. If there is insufficient force, the welding may not be of high quality. Figure 4a,c show the effects of peak current and frequency. As the peak current and frequency increase, the maximum load also increases. The gap between workpieces has a role in accelerating the outer workpiece during welding. Notably, the influence of the gap between workpieces was significant; it has optimized values compared to others for achieving the high-quality joint in Figure 4b.

Figure 5 provides a comparison of measured and calculated results obtained using the proposed prediction model for maximum load and weld length. The predicted results agree well with the experimental results. Errors between measured and calculated values were found to be less than 10%, indicating that the proposed prediction model accurately predicted maximum load and weld length.

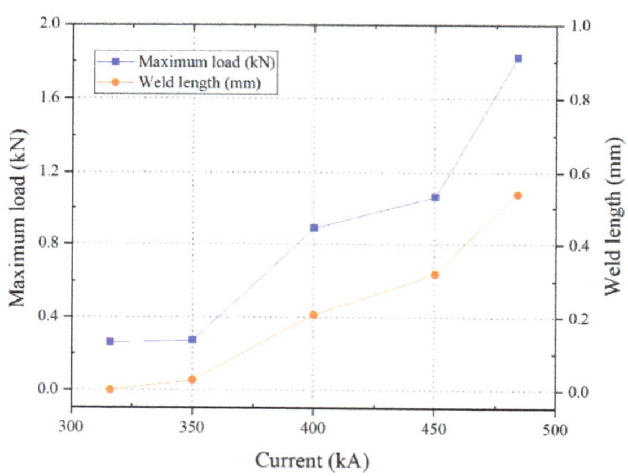

(**a**)

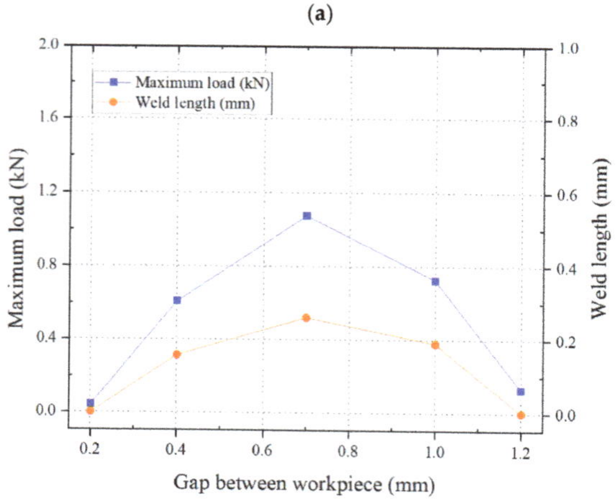

(**b**)

Figure 4. *Cont.*

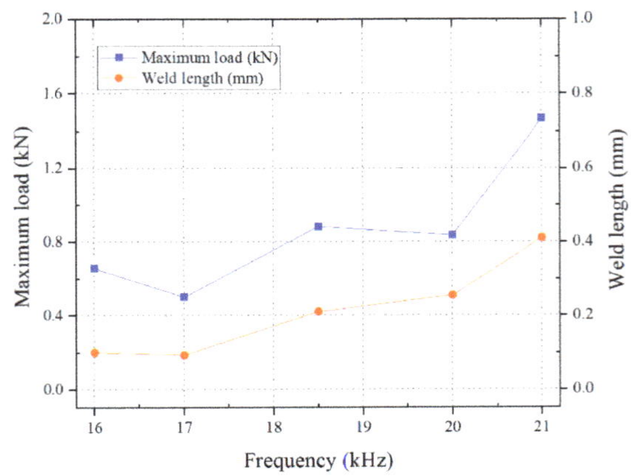

(c)

Figure 4. Effects of process parameters on Al/Steel joints. (**a**) Peak current. (**b**) Gap between workpieces. (**c**) Frequency.

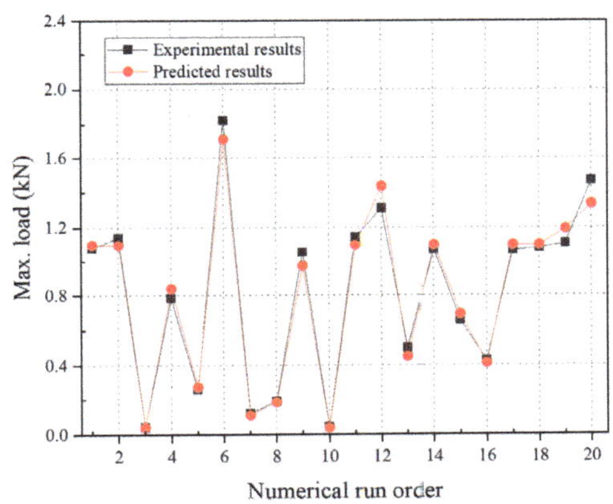

(a)

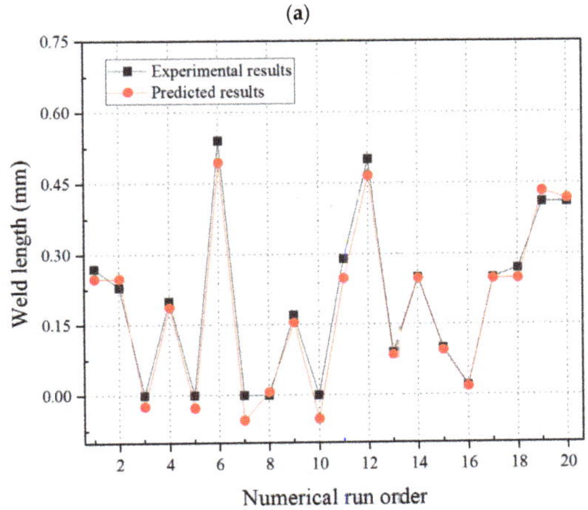

(b)

Figure 5. Comparisons between measured and calculated results. (**a**) Prediction accuracy of developed models for maximum load. (**b**) Prediction accuracy of developed models for weld length.

2.3.2. Optimization of Process Parameters

(1) Optimization by ICA

The optimization algorithm is a very useful method for finding the proper process parameters for various welding processes. One such algorithm is the Imperialist Competitive Algorithm (ICA), which was developed by Gargari et al. [30]. This is a new meta-heuristic optimization algorithm that incorporates political and social evolution [32]. The algorithm classifies the most powerful nations as imperialists and the rest as colonies. The process consists of five essential stages: empire initialization, assimilation, revolution, competition, and convergence. The first step, empire initialization, entails randomly selecting points from the function to initialize the empires. In an optimization problem with n dimensions, a country is represented by an n-dimensional array [28].

$$\text{Country} = [V_1,\ V_2,\ \cdots V_n] \tag{3}$$

$$\text{cost} = \text{F}(\text{country}) \tag{4}$$

The decision variables in this case are $V_1,\ V_2,\ \cdots V_n$ and the objective function is represented by F. The second is empire assimilation. The colonies that have a θ degree of deviation and x units move towards the imperialists. This deviation leads to a more thorough exploration of the decision space, as shown in Figure 6 [31]. θ and x are uniformly distributed random numbers. β is a number greater than one, d represents the distance between the imperialists and colonies, and γ is a parameter that determines the extent of deviation from the original direction [28,30–32].

$$\text{x} \sim \text{U}(0, \beta \times \text{d}) \tag{5}$$

$$\theta \sim \text{U}(-\gamma, \gamma) \tag{6}$$

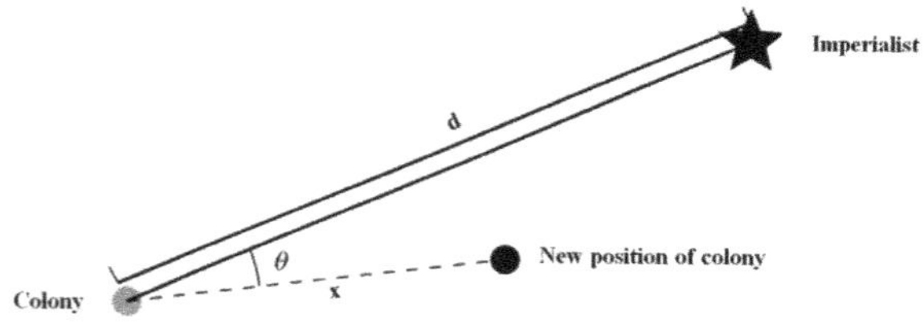

Figure 6. Progression of colonies towards imperialism [31].

As part of the empire-revolution process, certain colonies undergo random repositioning. Following this, in the fourth step, the weakest colony within the weakest empire group is targeted for pillaging by stronger empires as part of the empire competition. This occurrence is determined by the probability proportional to the strength of each imperialist power and the deviation from the average power of the colonies [29].

$$\text{T}.C_n = Cost(imperialist_n) + \xi mean\{Cost(Colonies\ of\ empire_n)\} \tag{7}$$

where $\text{T}.C_n$ is the total cost of the nth empire, and ξ is a positive number less than one.

Finally, the competition continues until there is only one imperialist, representing the strongest empire in the search space; this is known as empirical convergence [28–32].

In this study, the effects of varying the number of countries on the maximum load and weld length were investigated. Figure 7 shows the effects of this variation on joint quality. The objective function displayed notable variation when the number of countries was below 80. Increasing the number to 100 marginally enhanced the results. However, further increments in the initial number of countries resulted in decreases in the objective function, although these changes were not significant. Table 9 shows the optimum algorithmic parameters for the ICA model, including the number of countries. These parameters were determined by trial-and-error. The optimization models were developed using the MATLAB R2018b platform, and the developed prediction models in Section 2.3.1 were embedded into the ICA and optimized.

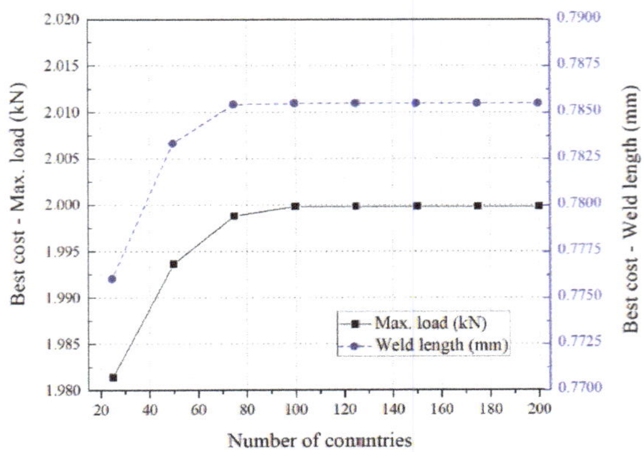

Figure 7. Effect of varying the number of countries on Al/Steel Joint.

Table 9. ICA parameters.

ICA Parameters	Al/Steel MPW
Revolution rate	0.3
Number of Countries	100
Number of Initial Imperialists	6
Number of decades	100
Assimilation Coefficient	1.5
Assimilation Angle Coefficient	0.5
Variable min (C, G, F) (kA, mm, kHz)	(316, 0.2, 16)
Variable max (C, G, F) (kA, mm, kHz)	(484, 1.2, 21)

Figures 8 and 9 show the average and minimum cost, indicating convergence to optimal values. The cost function exhibited a significant decrease in the first five decades. As the iteration proceeded beyond 35 decades, the variation in the cost function became minimal. Specifically, after 42 decades, the maximum load and weld length, which serve as objective functions in this study, were 2.00012 kN and 0.78892 mm, respectively.

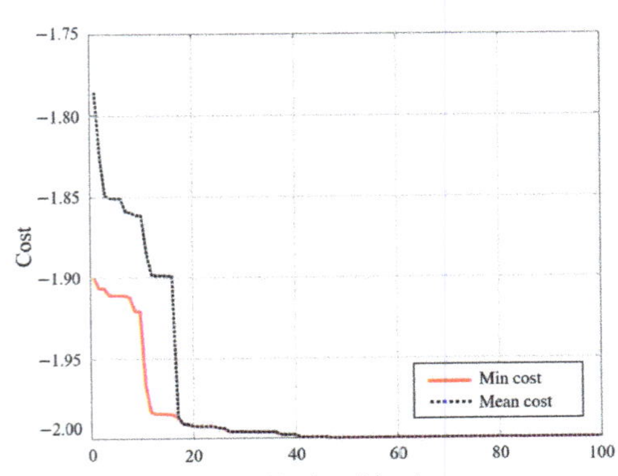

Figure 8. Convergence of maximum load objective function determined by ICA.

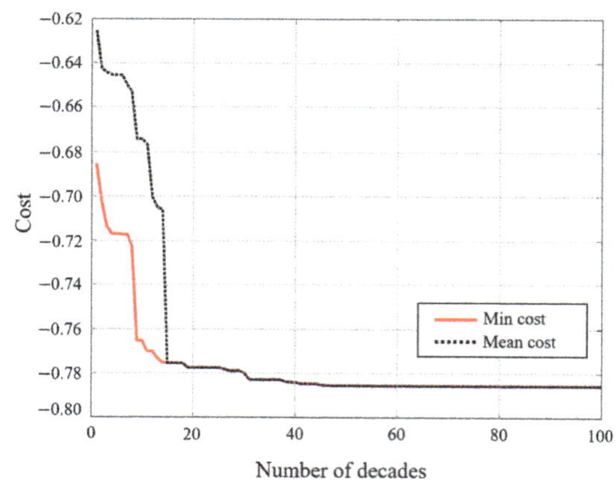

Figure 9. The convergence of the weld length objective function is determined by the ICA.

(2) Optimization by GA

GA, a popular optimization algorithm, is derived from natural evolution processes and originates from Darwin's theory of species evolution. J.H. Holland introduced GA in 1992 [26]. The algorithm utilizes chromosome representations, which are considered points in the solution space. The population's chromosomes are processed through genetic operators to successively replace them. Typically, a GA involves three key steps: selection, crossover, and mutation. First, a random population of chromosomes is created, and the fitness of each chromosome is assessed [24]. If the termination criterion is not met, the process moves on to the following steps. However, if the criterion is met, the best chromosome is returned. The creation of chromosomes continues until a total of n chromosomes are generated [25]. The initial population represents the possible solutions to the optimization problem, and each possible solution is referred to as an individual [23]. A possible solution was formed by the values of the peak current (kA), the gap between workpieces (mm), and the frequency (kHz). The search for the optimum was based on the maximization of the load and weld length, which define the objective function. Table 10 presents the GA parameters that yield the best results: a mutation rate of 0.008, a population size of 100, and a maximum number of generations of 100. Figures 10 and 11 depict the results achieved using the optimal process parameters obtained through GA. These parameters were determined in just 45 generations for the Al/Steel MPW joint. The maximum load and weld length, defining the objective function, were found to be 1.99929 kN and 0.78551 mm, respectively.

Table 10. GA parameters.

GA Parameters	Al/Steel MPW
Population size	100
Number of generations allowed	100
Mutation rate	100
Crossover rate	0.5
Type of crossover	0.5

(3) Comparison of optimization by ICA and GA

A systematic analysis was carried out to evaluate the effectiveness of the proposed algorithms. Both the ICA and GA algorithms were implemented in MATLAB and executed on an INTEL Core-i5 2500M laptop CPU. Table 11 presents a comparison of the optimization results obtained by ICA and GA. It is apparent that ICA slightly outperforms GA in terms of both accuracy and computational time.

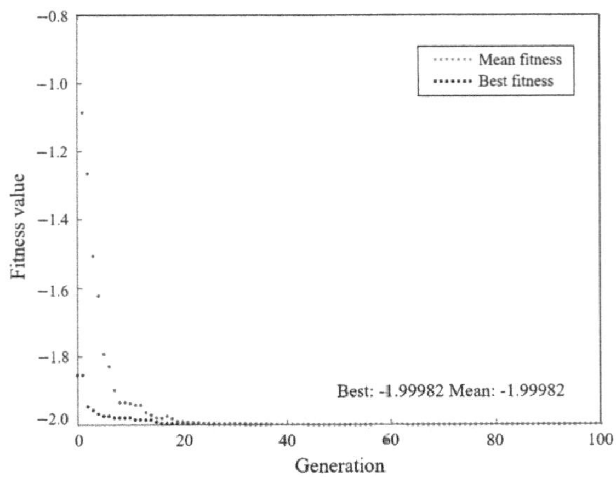

Figure 10. The convergence of the maximum load objective function is determined by GA.

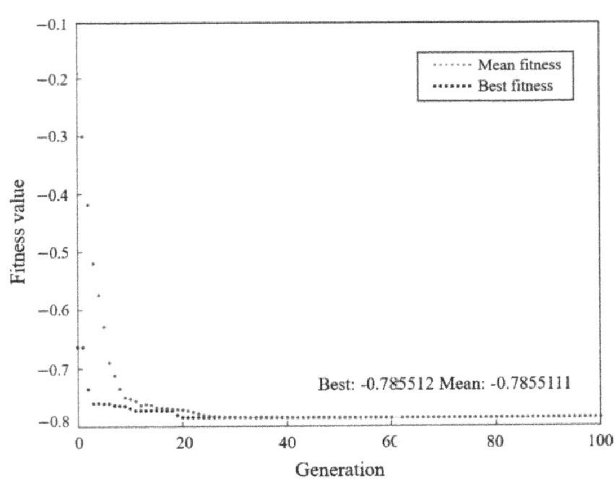

Figure 11. The convergence of the weld length objective function is determined by GA.

Table 11. Comparison between ICA and GA results.

Method	Model	Peak Current (kA)	Gap between Workpieces (mm)	Frequency (kHz)	Cost	CPU Time (s)
ICA	Maximum load (kN)	484	0.77498	21	2.00012	2.9
	Weld length (mm)	483.9	0.73533	21	0.78892	3
GA	Maximum load (kN)	483.9	0.77485	20.9	1.99929	8
	Weld length (mm)	483.9	0.73529	20.9	0.78551	9.4

The optimized process parameters were 484 kA, 0.73 mm, and 21 kHz for peak current, gap between workpieces, and frequency for achieving maximum weld length. Verification of the selected optimal process parameters was performed. The experiment used the optimized process parameters from Table 11 for the joining conditions. After welding, joint qualities were tested according to the maximum values of load and weld length. Figure 12 shows the results of the verification test. The joint at the Al tube was broken, indicating that the strength of the weld joints is greater than the strength of the weaker parent materials, as illustrated in Figure 12a. The maximum load was measured at 2.139 kN. The weld length was measured at 0.8 mm on the joint, as shown in Figure 12b.

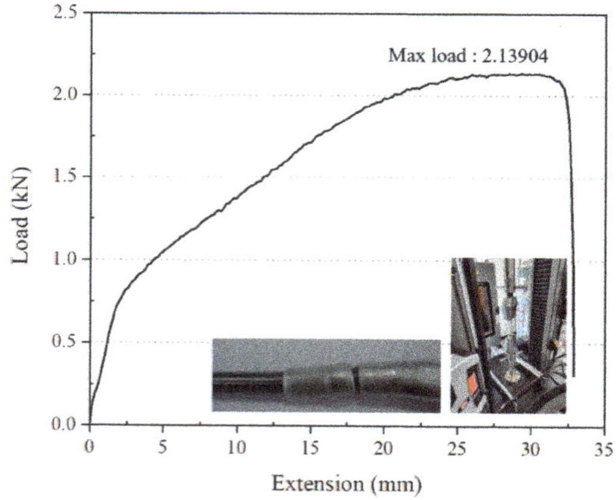

(**a**)

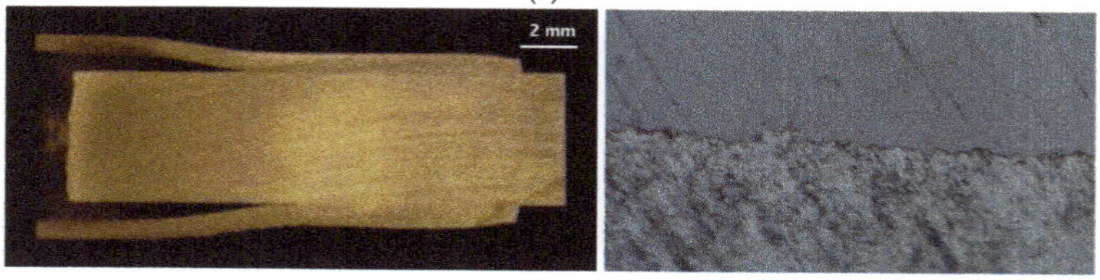

(**b**)

Figure 12. Results of the verification test. (**a**) Maximum load on Al/Steel joint. (**b**) Weld length on Al/Steel joint.

3. Conclusions

To improve the maximum load and weld length, this study used ICA and GA as evolutionary algorithms to determine optimal process parameters for Al/Steel MPW joints. The conclusions can be summarized as follows:

(1) The maximum load and weld length of Al/Steel joints varied under different welding conditions, as determined by CCD. In all cases, the maximum load was measured, but there were five cases in which no weld zone was observed, indicating a separation between Al tube and the steel rod during cutting. From these results, selected MPW process parameters were found to be closely related to joint quality.

(2) Based on the experimental results, prediction models for maximum load and weld length were developed using RSM. The developed prediction models showed high coefficients of determination (R^2) of 96.02% and 95.81% for maximum load and weld length, respectively. The deviation between actual response values and predicted values was less than 10%, indicating reasonable agreement between the developed prediction models and the experimental results.

(3) The peak current was discovered to be a crucial factor in enhancing joint quality, specifically in relation to maximum load and weld length, contributing over 48% to the overall quality. Increasing the peak current and frequency improved the quality because the peak current is closely related to the generation of the Lorentz force needed for the collision of workpieces during MPW.

(4) Optimization was performed using ICA and GA as evolutionary algorithms to select the optimal process parameters. Results from the proposed ICA and GA showed that

ICA performed at a slightly faster rate and yielded higher-quality results than GA. This indicates that the evolutionary algorithm was useful for optimizing the MPW process parameters.

(5) The optimal process parameters selected using ICA were verified through a verification test. The maximum load and weld length of the Al/Steel MPW joint were measured and found to be 2.139 kN and 0.8 mm, respectively. These results showed good agreement between the ICA-predicted and experimentally-obtained maximum values of load and weld length, confirming that ICA is an effective and useful method for finding optimal process parameters.

Author Contributions: Conceptualization, I.K. and J.S.; Methodology, J.S.; Software, J.S.; Formal analysis, J.S.; Investigation, J.S.; Resources, J.S.; Writing—original draft, J.S.; Writing—review & editing, I.K.; Supervision, I.K. All authors have read and agreed to the published version of the manuscript.

Funding: This research was supported by the Korea Institute of Industrial Technology (KITECH) under one of its major projects, "Development of Smart Welding System Modules with Dynamic Variable Control for Full Penetration Weld (KITECH EH-23-0007).

Data Availability Statement: The data presented in this study are available in this article.

Conflicts of Interest: The authors declare no conflict of interest.

References

1. Shim, J.Y.; Kim, I.S.; Kang, M.J.; Kim, I.J.; Lee, K.J.; Kang, B.Y. Joining of aluminum to steel pipe by magnetic pulse welding. *Mater. Trans.* **2011**, *52–55*, 999–1002. [CrossRef]
2. Kayode, O.; Akinlabi, E.T. An overview on joining of aluminum and magnesium alloys using friction stir welding (FSW) for automotive lightweight applications. *Mater. Res. Express.* **2019**, *6–11*, 112005. [CrossRef]
3. Zhao, D.; Liu, F.; Tan, Y.B.; Shi, W.; Xiang, S. Improving the strength-ductility synergy and corrosion resistance of Inconel 718/316L dissimilar laser beam welding joint via post-weld heat treatment. *J. Mater. Res. Technol.* **2023**, *26*, 71–87. [CrossRef]
4. Chen, L.; Wang, C.; Zhang, X.; Mi, G. Effect of parameters on microstructure and mechanical property of dissimilar joints between 316L stainless steel and GH909 alloy by laser welding. *J. Manuf. Process.* **2021**, *65*, 60. [CrossRef]
5. Pan, B.; Sun, H.; Shang, S.L.; Banu, M.; Wang, P.C.; Carlson, B.E.; Liu, Z.K.; Li, J. Understanding formation mechanisms of intermetallic compounds in dissimilar Al/steel joint processed by resistance spot welding. *J. Manuf. Process.* **2022**, *83*, 212–222. [CrossRef]
6. Yang, C.L.; Wu, C.S.; Lv, X.Q. Numerical analysis of mass transfer and material mixing in friction stir welding of aluminum/magnesium alloys. *J. Manuf. Process.* **2018**, *32*, 380–394. [CrossRef]
7. Bhattacharya, T.K.; Das, H.; Jana, S.S.; Pal, T.K. Numerical and experimental investigation of thermal history, material flow and mechanical properties of friction stir welded aluminium alloy to DHP copper dissimilar joint. *Int. J. Adv. Manuf. Technol.* **2017**, *88*, 847–861. [CrossRef]
8. Joshania, E.; Beidokhtib, B.; Davodib, A.; Amelzadeh, M. Evaluation of dissimilar 7075 aluminum/AISI 304 stainless steel joints using friction stir welding. *J. Alloys. Metall. Syst.* **2023**, *3*, 100017. [CrossRef]
9. Kumar, P.; Ghosh, S.K.; De, J.; Barma, S.; Saravanan, b.; Kumar, J. The joining of magnesium and aluminium alloys by inclined arrangement of explosive welding. *Mater. Today Proc.* **2023**, *76*, 536–541. [CrossRef]
10. Carvalho, G.H.S.F.L.; Mendes, R.; Leal, R.M.; Galvão, I.; Loureiro, A. Effect of the flyer material on the interface phenomena in aluminium and copper explosive welds. *Mater. Des.* **2017**, *122*, 172–183. [CrossRef]
11. Kapil, A.; Sharma, A. Comprehensive Weldability Criterion for Magnetic Pulse Welding of Dissimilar Materials. *Metals* **2022**, *12*, 1791. [CrossRef]
12. Faes, K.; Shotri, R.; De, A. Probing Magnetic Pulse Welding of Thin-Walled Tubes. *J. Manuf. Mater. Process.* **2020**, *4*, 118. [CrossRef]
13. Zhang, W.; Chen, Y.; Xie, J.; Zhang, T.; Wang, S.; Song, X.; Yin, L. Interfacial microstructure of Al/Ta dissimilar joints by magnetic pulse welding. *J. Mater. Res. Technol.* **2023**, *23*, 4167–4172. [CrossRef]
14. Geng, H.; Sun, L.; Li, G.; Cui, J.; Huang, L.; Xu, Z. Fatigue fracture properties of magnetic pulse welded dissimilar Al-Fe lap joints. *Int. J. Fatigue* **2019**, *121*, 146–154. [CrossRef]
15. Patra, S.; Singh, G.; Mandal, M.; Chakraborty, R.; Arora, K.S. Non-destructive evaluation and corrosion study of magnetic pulse welded Al and low C steel joints. *Mater. Chem. Phys.* **2023**, *309*, 128315. [CrossRef]
16. Yao, Y.; Chen, A.; Wang, F.; Jiang, H.; Li, G.; Cui, J. Mechanical properties and joining mechanisms of magnetic pulse welding joints of additively manufactured 316L and conventional AA5052 aluminum alloy. *J. Mater. Res. Technol.* **2023**, *26*, 6146–6161. [CrossRef]
17. Shim, J.Y.; Kang, B.Y.; Kim, I.S. Characteristics of Al/steel magnetic pulse tubular joint according to discharging time. *J. Mech. Sci. Technol.* **2017**, *31*, 3793–3801. [CrossRef]

18. Bembalge, O.B.; Singh, B.; Panigrahi, S.K. Magnetic pulse welding of AA6061 and AISI 1020 steel tubes: Numerical and experimental investigation. *J. Manuf. Process.* **2023**, *101*, 128–140. [CrossRef]
19. Yan, Z.; Xiao, A.; Cui, X.; Guo, Y.; Lin, Y.; Zhang, L.; Zhao, P. Magnetic pulse welding of aluminum to steel tubes using a field-shaper with multiple seams. *J. Manuf. Process.* **2021**, *65*, 214–227. [CrossRef]
20. Huang, M.L.; Hung, Y.H.; Yang, Z.S. Validation of a method using Taguchi, response surface, neural network, and genetic algorithm. *Measurement* **2016**, *94*, 284–294. [CrossRef]
21. Muhammad, W.; Husain, W.; Tauqir, A.; Wadood, A. Optimization of friction stir welding parameters of AA2014-T6 alloy using Taguchi statistical approach. *J. Weld. Join.* **2020**, *38–35*, 493–501. [CrossRef]
22. Ayaz, M.; Khandaei, M.; Vahidshad, Y. Investigating the effect of electromagnetic impact welding parameters on the microstructure evolution and mechanical properties of SS-Cu joint. *Mater. Today Commun.* **2023**, *35*, 105404. [CrossRef]
23. Liu, B.; Jin, W.; Lu, A.; Liu, K.; Wang, C.; Mi, G. Optimal design for dual laser beam butt welding process parameter using artificial neural networks and genetic algorithm for SUS316L austenitic stainless steel. *Opt. Laser Technol.* **2020**, *125*, 106027. [CrossRef]
24. Ye, G.Z.; Kang, D.K. Extended Evolutionary Algorithms with Stagnation-Based Extinction Protocol. *Appl. Sci.* **2021**, *11*, 3461. [CrossRef]
25. Li, Y.B.; Sang, H.B.; Xiong, X.; Li, Y.R. An Improved Adaptive Genetic Algorithm for Two-Dimensional Rectangular Packing Problem. *Appl. Sci.* **2021**, *11*, 413. [CrossRef]
26. Holland, J.H. *Adaptation in Natural and Artificial Systems*; The University of Michigan Press: Ann Arbor, MI, USA, 1975.
27. Katoch, S.; Chauhan, S.S.; Kumar, V. A review on genetic algorithm: Past, present, and future. *Multimed. Tools Appl.* **2021**, *80*, 8091–8126. [CrossRef] [PubMed]
28. Moghari, S.M.H.; Morovati, R.; Moghadas, M.; Araghinejad, S. Optimum Operation of Reservoir Using Two Evolutionary Algorithms: Imperialist Competitive Algorithm (ICA) and Cuckoo Optimization Algorithm (COA). *Water Resour. Manag.* **2015**, *29*, 3749–3769. [CrossRef]
29. Hemmati, A.; Bahrami, S.; Alaghebandha, M.; Ahmadifard, M. Solving Combined Model Inventory Control with Queuing Theory Approach Using Meta-Heuristic Algorithms. *Middle-East J. Sci. Res.* **2011**, *10–13*, 374–388.
30. Gargari, E.A.; Lucas, C. Imperialist Competitive Algorithm: An Algorithm for Optimization Inspired by Imperialistic Competition. In Proceedings of the 2007 IEEE Congress on Evolutionary Computation, Singapore, 25–28 September 2007; pp. 4661–4667. [CrossRef]
31. Wang, Z.S.; Lee, J.; Song, C.G.; Kim, S.J. Efficient Chaotic Imperialist Competitive Algorithm with Dropout Strategy for Global Optimization. *Symmetry* **2020**, *12*, 635. [CrossRef]
32. Luo, J.; Zhou, J.; Jiang, X. A Modification of the Imperialist Competitive Algorithm With Hybrid Methods for Constrained Optimization Problems. *IEEE Access* **2021**, *9*, 161745–161760. [CrossRef]
33. Madani, T.; Boukraa, M.; Aissani, M.; Chekifi, T.; Ziadi, A.; Zirari, M. Experimental investigation and numerical analysis using Taguchi and ANOVA methods for underwater friction stir welding of aluminium alloy 2017 process improvement. *Int. J. Press. Vessel. Pip.* **2023**, *201*, 104879. [CrossRef]

applied sciences

MDPI

Article

Effect of Temperature Distribution on Interfacial Bonding Process between CFRTP Composite and Aluminum Alloy during Laser Direct Joining

Qi Wang [1,2], Rao Fu [1,2], Fuji Wang [1,2,*], Chaoyang Luo [1,2], Jiankang Li [1,2] and Zhenyuan Jia [1,2]

1 State Key Laboratory of High-Performance Precision Manufacturing, Dalian University of Technology, Dalian 116024, China; wqwhh@mail.dlut.edu.cn (Q.W.); r.fu@dlut.edu.cn (R.F.); luochaoyang@mail.dlut.edu.cn (C.L.); 1323504160@mail.dlut.edu.cn (J.L.); jzyxy@dlut.edu.cn (Z.J.)

2 Key Laboratory of High-Performance Manufacturing for Advanced Composite Materials, Dalian 116024, China

* Correspondence: wfjsll@dlut.edu.cn

Abstract: Laser direct joining enables non-destructive and lightweight joining of carbon fiber reinforced thermoplastic (CFRTP) composites and aluminum alloys. The interfacial bonding process determines the joint performance and is influenced by the time-varying temperature distribution. However, the interfacial bonding process occurs inside the joint, making it difficult to study the effect of temperature distribution. To resolve this issue, a novel online observation device for the interfacial bonding process between CFRTP composites and aluminum alloys is design, and the polymer melting, flowing, and bonding with metal during laser direct joining are observed. Further, temperature field simulation models for laser direct joining are established, and temperature distribution and gradient are calculated. The results show that the temperature distribution determines the melting of CFRTP composites, and bubbles generated by the thermal decomposition of the polymer hinder the melting. The temperature gradient is related to the movement of the molten matrix and fibers, and the movement towards the aluminum alloy induces cracking and delamination. Once the interface is filled with polymer, the motion changes to along the laser scanning direction and the joining defects are reduced. The study can provide a foundation for promoting interfacial bonding and reducing the defects of laser direct joining.

Keywords: laser direct joining; carbon fiber reinforced thermoplastic composite; aluminum alloy; interfacial bonding process

check for updates

Citation: Wang, Q.; Fu, R.; Wang, F.; Luo, C.; Li, J.; Jia, Z. Effect of Temperature Distribution on Interfacial Bonding Process between CFRTP Composite and Aluminum Alloy during Laser Direct Joining. *Appl. Sci.* **2023**, *13*, 11973. https://doi.org/10.3390/app132111973

Academic Editors: Rui Manuel Leal, Gustavo H. S. F. L. Carvalho and Ivan Galvão

Received: 11 October 2023
Revised: 28 October 2023
Accepted: 30 October 2023
Published: 2 November 2023

1. Introduction

Lightweight design and manufacturing are the development direction of aerospace, transportation, and other fields, which can realize performance improvement, energy saving, and emission reduction [1,2]. Carbon fiber reinforced polymer composites have the advantages of high specific strength, high specific stiffness, designable properties, and integral fabrication of components, making them the preferred materials for reducing weight and improving performance [3,4]. Depending on the polymer matrix, such composites can be divided into thermosetting and thermoplastic. Both types of composites need to be used in applications with metals such as aluminum and titanium alloys, thus weakening the negative effects of strong anisotropy, low interlaminar shear properties, and the high manufacturing costs of composites [5,6]. For instance, the fuselage of the Airbus A350XWB airliner is mainly made of carbon fiber reinforced polymer composites, while aluminum and titanium alloys are used for the reinforcement of certain high-load-bearing structures [7,8].

Compared with thermoset composites, carbon fiber reinforced thermoplastic (CFRTP) composites exhibit stronger impact resistance, as well as the advantages of heat molding, fusion joining, and recycling [9–11]. With the increasing maturity of raw material preparation and component molding, they are gradually used in aircraft flat tails, fuselages, etc., and face the same problem of joining with metals [12–14]. Mechanical joining methods such as bolting and riveting are commonly used for structures subjected to concentrated loads. For structures subjected to uniform or shear loads, fusion joining, which relies on interfacial bonding formed by molten thermoplastic polymers, is gaining importance. This approach avoids the performance degradation of CFRTP composites caused by material processing and reduces the use of adhesives and fasteners, thereby maximizing the effect of weight reduction [15,16]. Depending on the heat source, fusion joining of CFRTP composites with metals can be categorized as friction joining, ultrasonic joining, laser joining, induction joining, and resistance joining. Among them, laser direct joining is the approach where a high energy density laser beam is used to heat the metal, putting the polymer matrix in contact with metal to melt and bond with metal to form a joint, with the advantages of good adaptability to material types and geometrical structures and easy automation [17–20].

The joining temperature is the key to determining whether CFRTP composites and metals can be joined through fusion [21,22]. Scholars have investigated the effect of joining temperature on the laser direct joining process using simulations and experiments. Hussein et al. equated the laser energy input to a Gaussian surface heat source and simulated the temperature field when laser direct joining PMMA and stainless steel [23]. To improve the accuracy of the temperature simulation, the temperature-dependent thermal properties and the anisotropy of the CFRTP composite were further considered, which allowed the average error between the simulated and measured temperatures to be less than 10% [24,25]. Based on the temperature field simulation during laser direct joining, scholars investigated the relationship between temperature distribution and the melting and solidification process of CFRTP composites. Jiao et al. used the melting point of the polymer matrix as the melting boundary criterion in the temperature simulation, that is, 280 °C for the polyphenylene sulfide used, and found that the melting widths and depths obtained from the simulation agreed with the experiment [26]. Similarly, Zhang et al. used a melting temperature of 303 °C for a nylon 6T matrix as the melting boundary criterion, and the CFRTP composite solidification sequence during the laser direct joining was investigated [27]. Further, considering the effect of melt flow on temperature contributes to the accurate determination of the melting boundary, Ai et al. developed a simulation model of laser joining considering the melting and flow processes and found that thermal convection in the molten polymer increased the melting width, and increasing the laser power increased the rate of thermal convection [28].

The joining temperature is not only related to the melting and solidification at the interface, but also determines the defect formation during the laser direct joining. Tao et al. [29] measured the interfacial temperature between a CFRTP composite and titanium alloy using a K-type thermocouple, and Lambiase et al. [30] measured the surface temperature of an aluminum alloy during laser direct joining using an infrared thermographic camera. They both found insufficient melting of the CFRTP composite at low temperatures and bubbles from thermal decomposition at high temperatures. Based on these studies, it was proven that laser direct joining can be improved by the interfacial temperature control, which ranges between the melting and thermal decomposition temperatures of the polymer matrix [31]. Meanwhile, it is also important to increase the interfacial bonding area of the CFRTP composite to the metal. However, in optimized laser direct joining processes based on temperature control, the bonding area is related to the melting width and can be increased by increasing the temperature, but is accompanied by the risk of thermal decomposition of the polymer [32]. Considering that thermal convection in the molten polymer contributes to the increase in the melting width, to simultaneously promote interfacial bonding and reduce joining defects, it is also necessary to investigate the effect of temperature distribution on the polymer matrix melting, flow, and bonding with the metal.

To reveal the effect of temperature distribution on interfacial bonding between CFRTP composites and metals, this paper designs an observation device for the interfacial bonding process during laser direct joining, and the carbon fiber reinforced polyether-ether-ether-ketone thermoplastic composite (CF/PEEK) and 6061 aluminum alloy used in aviation are taken as the experimental materials, realizing the observation of the polymer melting, flowing, and bonding with metal during laser direct joining. For the interface with and without polymer layer filling, simulation models of the joining temperature field of considering the anisotropic heat transfer and temperature-varying material properties are established, respectively. The relationship between the joining temperature and the melting and flow of the thermoplastic polymer is analyzed, and the influence of the filled polymer on the interfacial bonding process is revealed, which can provide a foundation for promoting interfacial bonding and reducing the defects of laser direct joining.

2. Materials and Methods

2.1. Experimental Device and Setup for Interfacial Bonding Process Observation

To observe the interfacial bonding process during laser direct joining of CFRTP composite with aluminum alloy, an online observation experimental device as shown in Figure 1 was designed. Principle of observation of interfacial bonding process: Firstly, the polished end faces of the CFRTP composite and the aluminum alloy were pressed onto a refractory glass, thus equating them to the confined material interface during laser direct joining; subsequently, a laser was used to irradiate the aluminum alloy close to the refractory glass, and a camera with a microscope was used to observe the interfacial bonding process through the refractory glass. The observation device consisted of the fixture, refractory glass, microscope, camera, focus adjuster, height adjuster, and laser baffle. The camera with microscope had a spatial resolution of 1.23 μm/pixel and a frame rate of about 40 Hz, satisfying the observation requirements of the interfacial bonding process. To accurately observe the interfacial bonding process, thermal insulation was made at the contact area between the fixture and the materials.

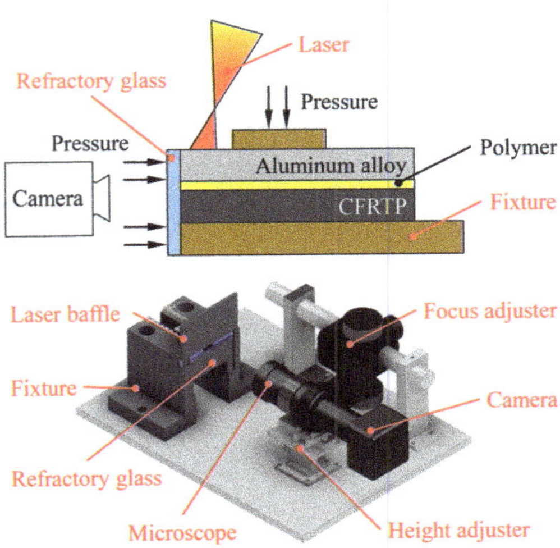

Figure 1. Observational principle and device for interfacial bonding process during laser direct joining of CFRTP composite with aluminum alloy.

The experimental materials were carbon fiber reinforced polyether-ether-ether-ketone (CF/PEEK) thermoplastic composite, 6061 aluminum alloy, and polyether-ether-ether-ketone (PEEK) polymer, which have been used in aviation and other fields. The lengths and widths were 100 mm × 25 mm, and the thicknesses were 2 mm, 3 mm, and 0.25 mm, respectively. The mechanical and thermal properties of the experimental materials are shown in Table 1. The CFRTP composite consisted of 16 plies alternating at 0° and 90°, and its fiber volume fraction was 60%. As shown in Figure 2, the laser direct joining experimental device consisted of a laser head, motion platform, observation device, and acquisition computer. Since the quality of the end faces of the materials directly determines the effect of observation, 120#, 400#, 1000#, and 3000# sandpapers were used to polish the end faces of the CFRTP composite and aluminum alloy until the end faces were smooth and without obvious scratches. Once the materials were prefixed to the fixture, it was necessary to apply pressure on the refractory glass. After ensuring that the polished end faces of the materials were in close contact with the refractory glass, the materials were clamped under the same torque using a torque wrench, and the observing position was adjusted to the center of the interface using the height and focus adjusters. Finally, to prevent the laser scattering from affecting the observation effect, the laser baffle was attached to the fixture.

Table 1. Mechanical and thermal properties of the experimental materials [25].

Parameters	CF/PEEK	6061 Aluminum Alloy	PEEK Polymer
Density (kg/m^3)	1580	2700	1300
Tensile strength (MPa)	980	320	100
Melting temperature (°C)	343	615–655	343
Decomposition temperature (°C)	520	-	520
Specific heat capacity J/(kg·K)	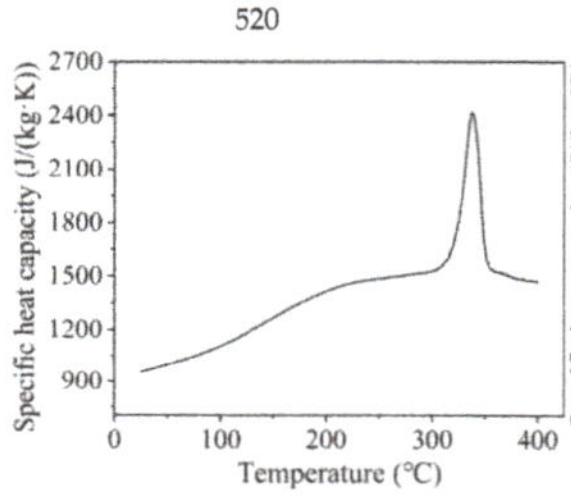	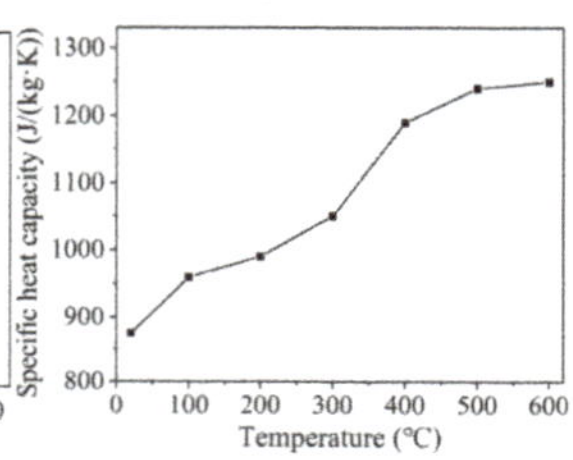	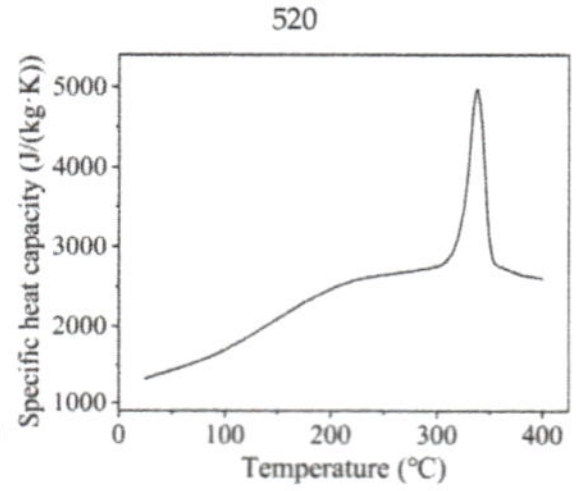
Thermal conductivity W/(m·K)	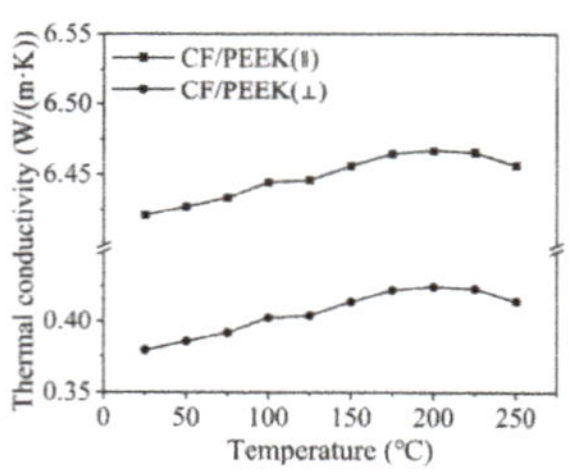	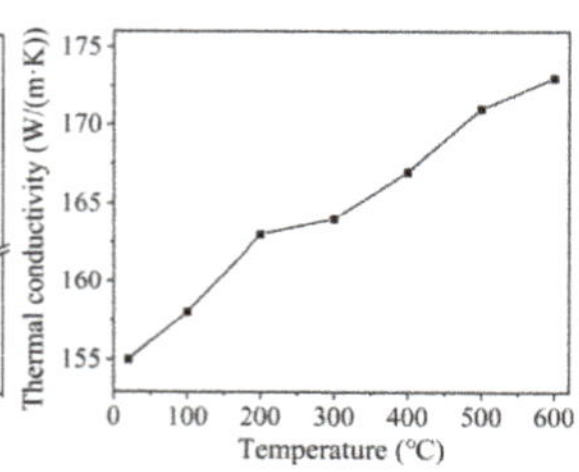	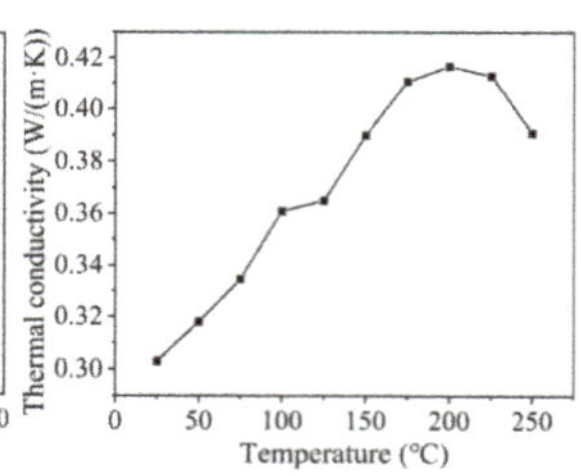

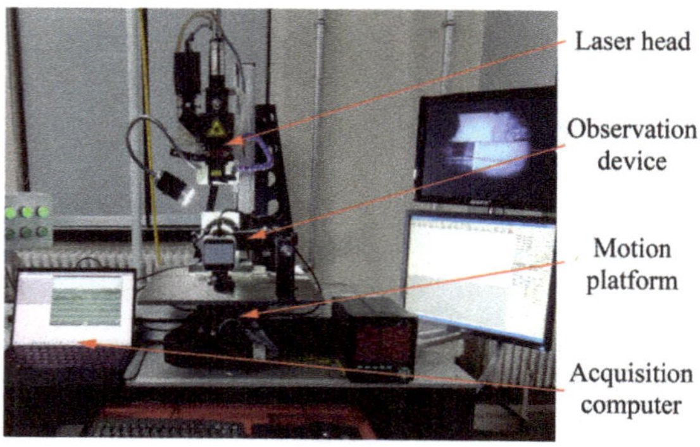

Figure 2. Laser direct joining experimental device.

Because of the high fiber volume fraction of the CFRTP composite, insufficient polymer matrix may lead to poor interfacial bonding during laser direct joining, and filling the polymer layer at the interface between the CFRTP composite and the aluminum alloy is a common solution [33]. To this end, laser direct joining experiments with and without polymer layer filling were performed as shown in Figure 3. In the experiments, a fiber laser with a wavelength of 1064 nm was used to scan the aluminum alloy surface reciprocally six times. The laser beam was in defocused state; the defocusing distance was 80 mm, the corresponding spot diameter was 5 mm, the maximum laser power was 1000 W, and the maximum laser energy density was about 5×10^3 W/cm^2. For both cases, the scanning speed of the laser was 10 mm/s, the single scanning length was 20 mm, and the center of the laser spot was 2.5 mm from the edge of the aluminum alloy. When there was no filled polymer layer, the laser powers were 400 W, 500 W, and 600 W, whereas the laser powers were 500 W, 600 W, and 700 W when the interface was filled with polymer layer, because the polymer has a high specific heat capacity and low thermal conductivity.

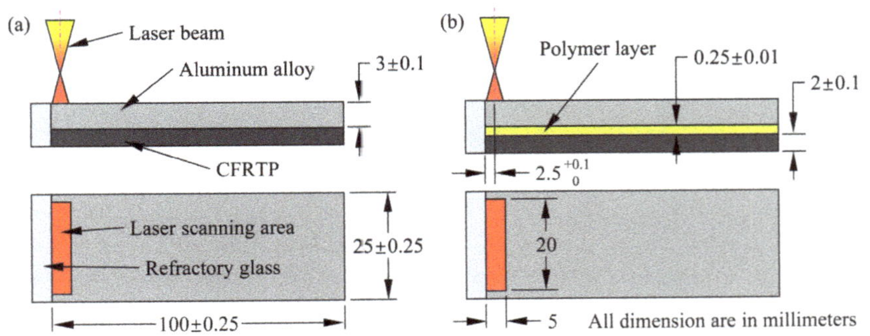

Figure 3. Schematic of laser direct joining of CFRTP composite with aluminum alloy: (**a**) interface without polymer filling and (**b**) with polymer filling.

2.2. Simulation Modeling of Temperature Field during Laser Direct Joining

To simulate the temperature field during laser direct joining, the geometrical models shown in Figure 4 were established for the interface with and without polymer layer filling, respectively, and the material dimensions were the same as those in Section 2.1. To balance the computational accuracy and efficiency, two sizes of grids were chosen for meshing. Since the temperature gradient in the heated region Γ_1 irradiated by the laser was high, a fine hexahedron with a size of 0.625 mm was used for meshing parallel to the thickness direction. The other region, Γ_2, which had a low temperature gradient, was meshed with a coarse hexahedral of size 2.5 mm, and wedges were used for the transition between the two regions. In the material thickness direction, aluminum alloy was meshed using 0.5 mm, and the polymer layer was meshed using 0.25 mm. Considering the anisotropy and ply structure of the CFRTP composite, it was meshed using ply thickness of 0.125 mm.

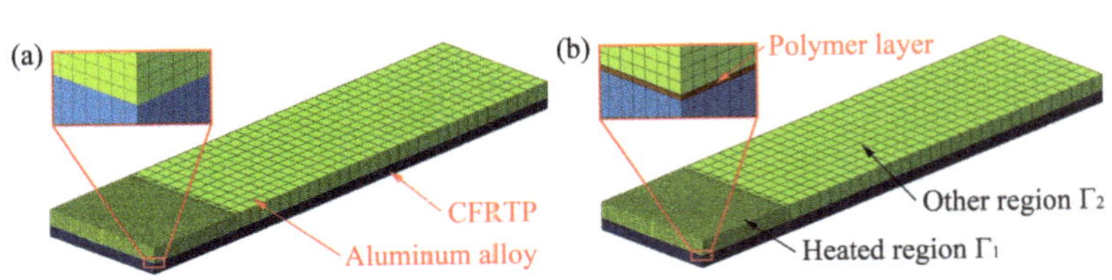

Figure 4. Meshed geometrical models for laser direct joining of CFRTP composite to aluminum alloy: (**a**) interface without polymer filling and (**b**) with polymer filling.

After establishing the meshed geometric models, it is necessary to set the governing equations and boundary conditions of the laser direct joining, including the material properties, the laser heat source, and the cooling conditions. Specific details can be found in our previous study [25] and are briefly described here. The temperature field of the laser direct joining obeys the heat transfer equation in (1):

$$\rho C_p \frac{\partial T}{\partial t} = k_x \frac{\partial^2 T}{\partial x^2} + k_y \frac{\partial^2 T}{\partial y^2} + k_z \frac{\partial^2 T}{\partial z^2} \tag{1}$$

where ρ is the density; C_p is the specific heat capacity; T is the temperature; t is the time; and k_x, k_y, and k_z are the thermal conductivity in the x, y, and z directions, respectively. To solve the above heat transfer equation, the initial temperature T_0 was set to be 20 °C, and the material properties were set based on Table 1. Additionally, based on the laser heat input and convective and radiative heat transfer on the outer surface of the material, the boundary conditions for the heated region Γ_1 and other regions Γ_2 were set as follows:

$$k\frac{\partial T}{\partial n} = h(T_s - T_0) + \varepsilon\sigma(T_s^4 - T_0^4) + q \text{ on } \Gamma_1 \tag{2}$$

$$k\frac{\partial T}{\partial n} = h(T_s - T_0) + \varepsilon\sigma(T_s^4 - T_0^4) \text{ on } \Gamma_2 \tag{3}$$

where n is the normal vector of the outer surface; q is the laser heat flow density; h is the convective heat transfer coefficient, which was set to 5 $W/(m^2{\cdot}K)$; T_s is the surface temperature; ε is the emissivity; and σ is the Stefan Boltzmann constant, which was 5.67×10^{-8} $W/(m^2{\cdot}K^4)$). The emissivity coefficients of CFRTP composite and aluminum alloy were roughly set because of the little effect of radiative heat transfer on the temperature during the laser direct joining [34,35]. Since the laser energy density was about 10^3~10^4 W/cm^2, and the aluminum alloy did not melt or had a small melting thickness, the Gaussian surface heat source was used to equalize the laser energy input. The expression is as follows:

$$q = \frac{3\eta P}{\pi R^2} \exp(-\frac{3[(x - x_0 - vt)^2 + (y - y_0)^2]}{R^2}) \tag{4}$$

where η is the laser absorption rate of the aluminum alloy; P is the laser power; R is the effective radius of the laser, that is, the radius corresponding to when the laser intensity is attenuated to 5% of the maximum value; x_0 and y_0 are the initial irradiation position of the laser; and v is the laser scanning speed. Among them, the laser absorption rate of aluminum alloy η was set to 0.3 based on the inverse analysis method reported by Lambiase et al. [34], the effective radius R of the laser was set to 2.5 mm, and the corresponding defocusing distance of the laser was 80 mm.

For the heat transfer at the material interface, since the polymer matrix of the CFRTP composite experiences the heating, melting, and flowing processes, the contact thermal resistance only influences the temperature field before the melting. On the other hand, contact thermal resistance is affected by external pressure, which is required for laser direct joining of CFRTP composite with aluminum alloy; Wang et al. have shown that when the external pressure reached a certain level it could be regarded as an ideal contact heat transfer [36]. Therefore, the effect of contact thermal resistance was ignored during the temperature simulation of laser direct joining. The temperature field during laser direct joining was solved using ABAQUS, and transient thermal analysis was performed on a simulation platform with an i5-11500 CPU and 8 GB of RAM.

3. Results and Discussion

3.1. Consistency Verification of Observation and Simulation Results

To verify the relationship between the temperature distribution and the melting of the CFRTP composite, the observed interfacial bonding processes and the corresponding temperature distributions at the material end faces in the simulation were compared as shown in Figures 5 and 6. In the observed results, the melting of the polymer matrix caused the thickness change in the CFRTP composite, and the melting depth of the CFRTP composite was difficult to accurately determine; therefore, the number of melted plies of CFRTP composite was used to compare the simulated and observed results. In the simulation results, the melting temperature (343 °C) of the PEEK polymer was used as the melting boundary criterion, and the corresponding melting area was indicated in gray. When the interface was not filled with polymer, Figure 5a shows the results at a laser power of 400 W. The position change of the fibers in the CFRTP composite and the cracking between the matrix and fibers were observed in the first ply, whereas the other plies were unchanged, indicating that melting only occurred in the first ply. Continuing to increase the laser power to 500 W and 600 W, the melting of the three and six plies, as shown in Figure 5b,c, respectively, was found in the observed results using a similar criterion. The observed number of melted plies was the same as the number of melted plies in the gray area in the simulation, indicating that there was a correlation between the number of melted plies and the temperature distribution. With the increase in laser power, the time corresponding to the maximum number of melted plies was delayed because the temperature of the aluminum alloy was higher at higher laser powers, and the CFRTP composite was still heated by a high-temperature aluminum alloy after the laser stopped irradiating the observation area. However, when the number of melted plies increased, the fibers in the direction of the parallel end faces were extruded, which affected the observation of the interfacial bonding process.

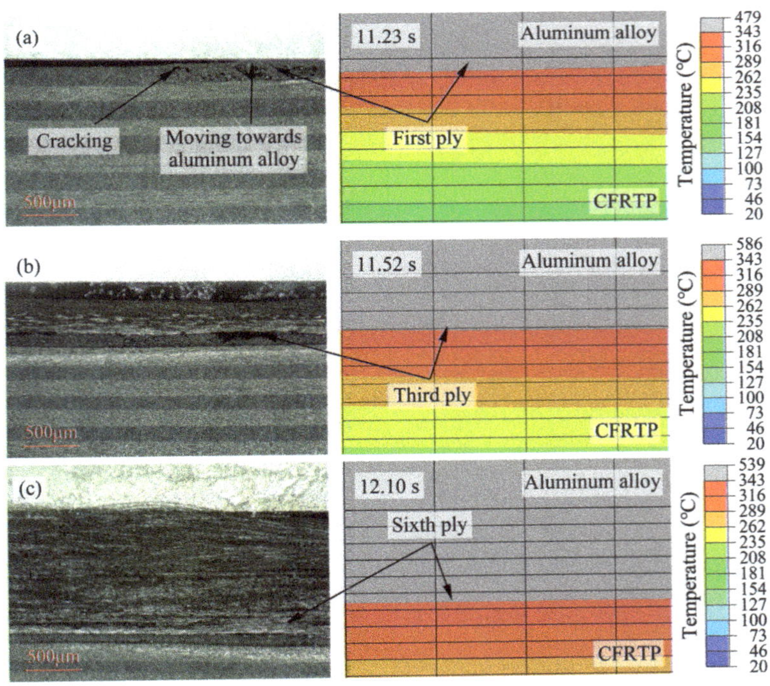

Figure 5. Observed interfacial bonding processes and the corresponding temperature distributions when the interface was not filled with polymer: (**a**) laser power 400 W, (**b**) 500 W and (**c**) 600 W.

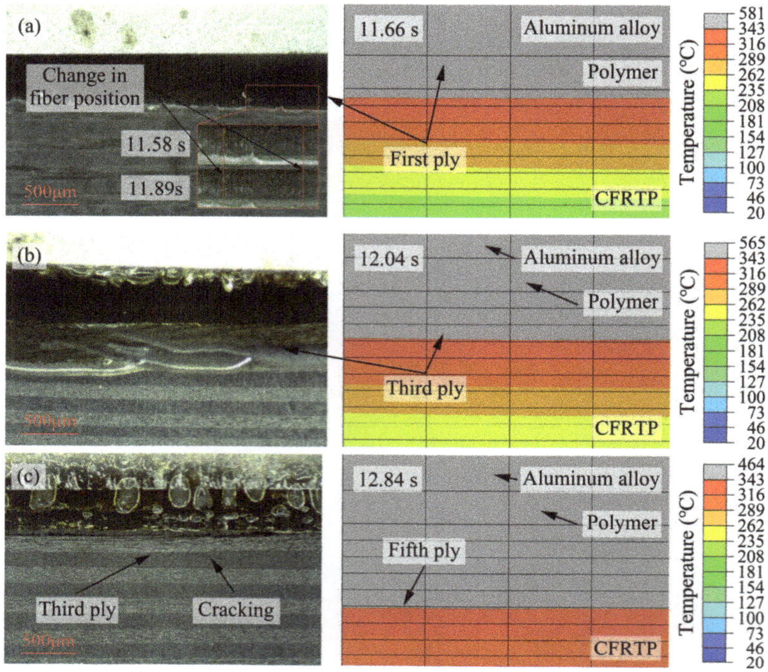

Figure 6. Observed interfacial bonding processes and the corresponding temperature distributions when the interface was filled with polymer: (**a**) laser power 500 W, (**b**) 600 W, and (**c**) 700 W.

The observed results and corresponding temperature distributions after filling the polymer layer at the interface are shown in Figure 6. The molten polymer layer filled the gap between the end face of the material and the refractory glass, leading to a more pronounced color change compared with the melting. However, this did not mean that melting of the polymer matrix occurred, and it was still necessary to use the position change of the fibers or cracking between the matrix and fibers as the melting criterion. Due to the lower thermal conductivity of the polymer than the CFRTP composite, the number of melted plies was lower under the same laser power. When the laser power was 500 W, cracking was not observed in Figure 6a, but the observations at 11.58 s and 11.89 s showed that the position of the fibers in the first ply changed, indicating that the polymer layer was completely melted and the first ply was partially melted, which was consistent with the simulation. With similar criterion, it was found that the number of melted plies in the CFRTP composite increased to three plies when the laser power was 600 W, and it was accompanied by thermal decomposition in the molten polymer layer. When the laser power was 700 W, the number of melted plies in the CFRTP composite was still observed to be three plies, which was different from the simulation result of five plies, because bubbles in molten polymer induced by the thermal decomposition hindered the heat transfer. The above results proved that the temperature distribution during laser direct joining was correlated with the melting of the polymer matrix and the interfacial filled polymer.

To investigate the effect of bubbles formed by the thermal decomposition of molten polymer on the bonding process, the results of interfacial observations and temperature simulations were extracted for the laser powers of 600 W, shown in Figure 7, and 700 W shown in Figure 8, where the gray area in the temperature distribution was the area beyond the thermal decomposition temperature of the polymer (520 °C). When the laser power was 600 W, the surface temperature of the polymer layer exceeded the thermal decomposition temperature, and bubbles were observed on the corresponding surface of the polymer during the last laser scan. The temperature simulation results showed that the local temperature of the polymer surface exceeded the thermal decomposition temperature at 10.75 s, and the peak temperature of the polymer surface decreased as the laser scanned from the edge to the center. However, from the observations of the corresponding moments, it was found that the bubbles moved toward the laser scanning direction, resulting in the presence of bubbles in all the moments, which may be related to the thermal convection caused by the uneven temperature. Meanwhile, similar fiber position changes were found in the CFRTP composite, but the fiber position change was smaller than the bubbles in the molten polymer because of the constraints from the fiber axial direction. When the laser power was 700 W, as shown in Figure 8, the surface temperature of the thermoplastic polymer exceeded the thermal decomposition temperature at 9.00 s, and bubbles appeared not only on the surface of the molten polymer, but also in the contact area with the CFRTP composite. Since the thermal conductivity of the gas in bubbles was significantly lower than that of the molten polymer, the bubbles reduced the heat transfer from the polymer to the CFRTP composite, which on the one hand made it easier to generate high temperatures in the molten polymer and exacerbated thermal degradation, and on the other hand reduced the number of melted plies of the CFRTP composite, which resulted in a lower number of melted plies than that of the simulation results.

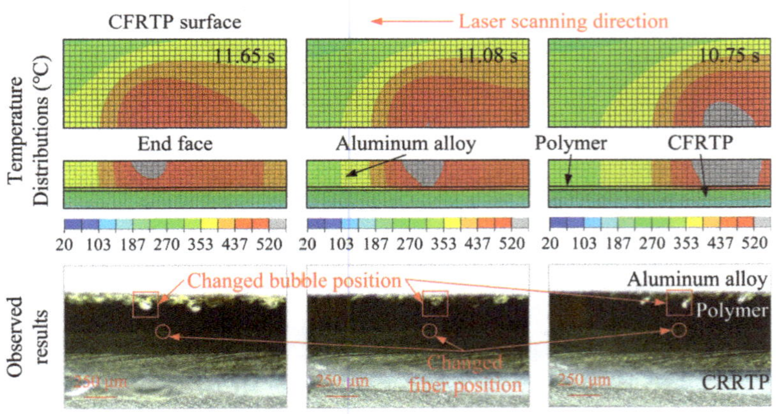

Figure 7. Simulated temperature distributions and corresponding observed interfacial bonding processes at different moments when the laser power was 600 W.

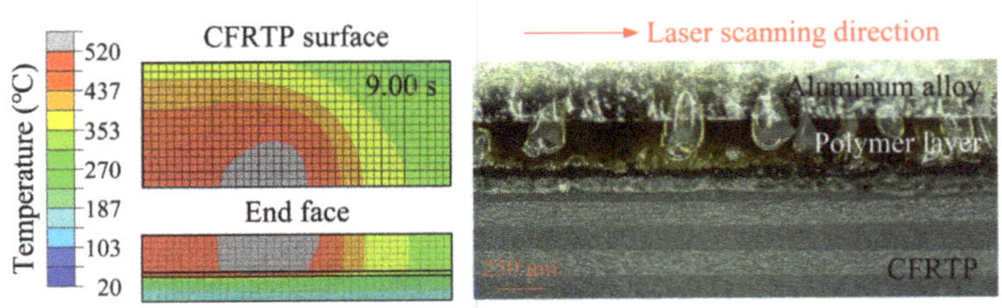

Figure 8. Simulated temperature distributions and corresponding observed interfacial bonding process at 9.00 s when the laser power was 700 W.

3.2. Effect of Temperature Distribution on Interfacial Bonding Process

To clarify the effect of temperature distribution on the interfacial bonding process, the end face temperatures at different moments when the laser power was 600 W were extracted, and the corresponding temperature gradients were calculated using MATLAB, with the results shown in Table 2. For the interface with and without polymer filling, the temperature distributions in the aluminum alloy were similar, that is, the temperature at the center of the laser irradiation was the highest, a trailing appeared along the laser scanning direction, and the temperatures were uniformly distributed in the thickness direction. In contrast, the temperatures in the CFRTP composite and filled polymer showed a significant decrease along the thickness direction, and the filled polymer resulted in more temperature decreases. The maximum temperature gradient in the aluminum alloy was about 45 °C/mm, whereas that of the polymer layer was about 450 °C/mm and larger than that of the unfilled polymer of about 420 °C/mm. For the x and y direction temperature gradients, the y direction temperature gradient was consistent with the temperature gradient, while the x direction temperature gradient was related to the laser scanning direction. The x direction temperature gradient in the aluminum alloy was negative in the laser scanned area and positive in the unscanned area, when the laser scanned in the $x-$direction, with maximum and minimum temperature gradients occurring on the sides of the laser irradiation position. The x direction temperature gradients in the CFRTP composite and filled polymers were similar, with reduced positive and negative peak gradients and a trailing near the contact area with the aluminum alloy, which was mainly related to the heating hysteresis caused by the low thermal conductivity.

Table 2. Temperature distributions and gradients at the end face when the laser power was 600 W.

	Interface without Polymer Filling	Interface with Polymer Filling
Temperature distribution	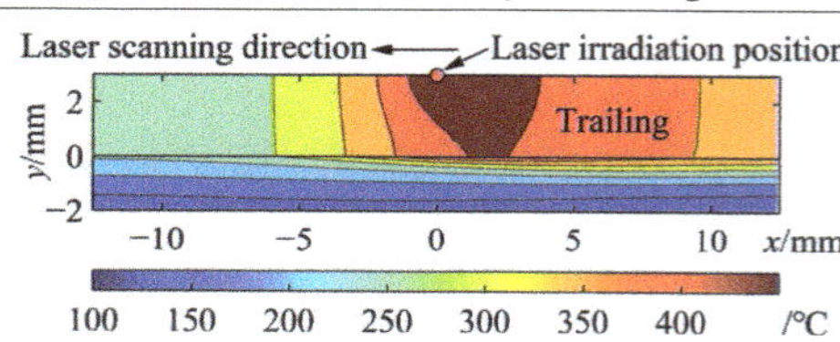	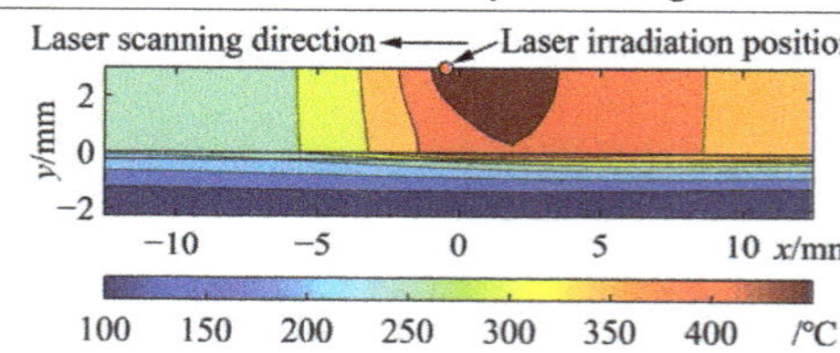
Temperature gradient	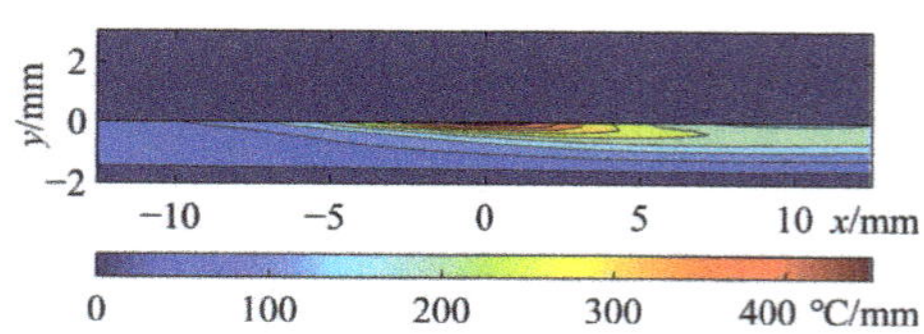	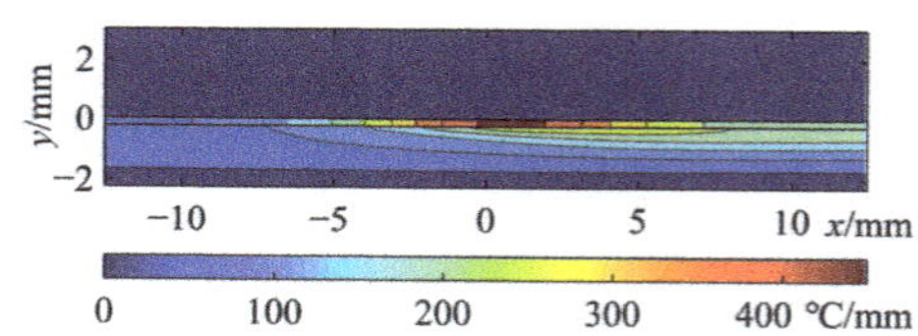
x direction temperature gradient	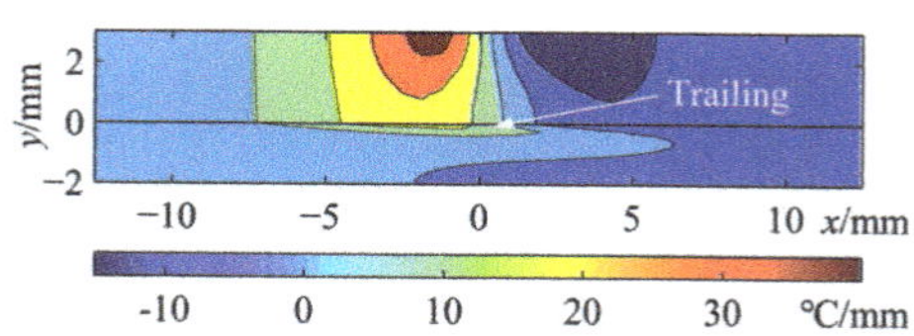	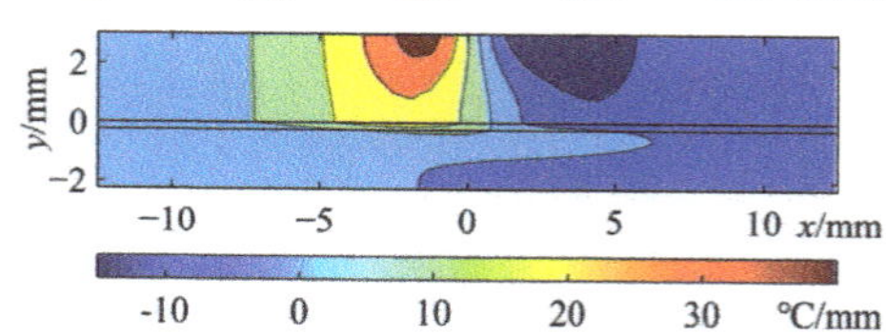
y direction temperature gradient	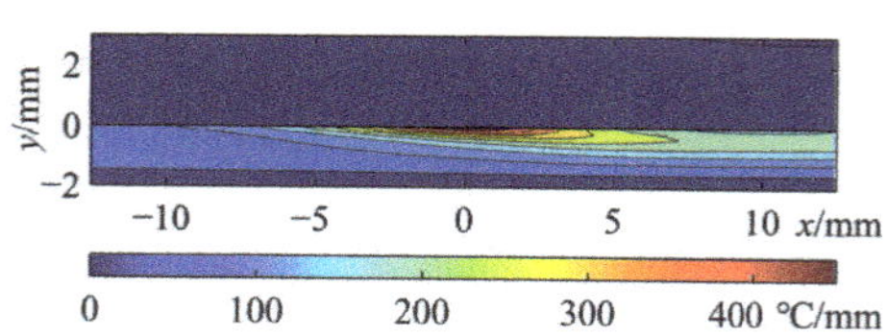	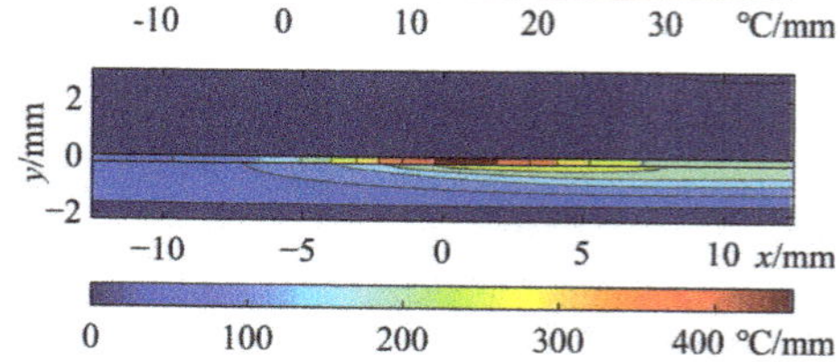

When the interface was not filled with polymer, Figure 9 shows the interfacial bonding process between the CFRTP composite and the aluminum alloy at a laser power of 500 W. The laser scanning direction is from right to left, and the laser irradiated the surface of the aluminum alloy in the center of the observation area at 7.00 s. At this moment, a gap of about 60 μm existed at the interface between the CFRTP composite and the aluminum alloy on the left side, while the gap on the right side was filled by the CFRTP composite. As the laser scanned to the left, the fibers in the CFRTP composite moved towards the aluminum alloy driven by the molten polymer matrix and filled the gap in the observed area by 7.19 s. Meanwhile, this also produced cracking between the matrix and fibers and formed the delamination between the first and second plies of the CFRTP composite. Similarly, the melting of the polymer matrix in the second ply was observed at 7.24 s, and the movement of the second ply towards the aluminum alloy eliminated the cracking in the first ply, but this likewise produced such defects in the second ply. In addition to the movement of the plies in the CFRTP composite towards the aluminum alloy, movement of the molten polymer matrix together with the fibers towards the laser scanning direction was also found, based on the position change of the cracking in the observations of the material end face from 7.14 s to 7.43 s.

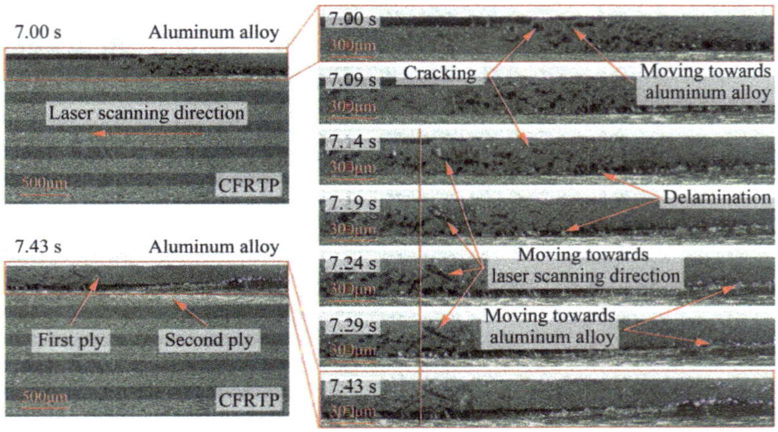

Figure 9. Interfacial bonding processes at different moments when the polymer was unfilled.

To determine the effects on the interfacial bonding process, the x and y direction temperature gradients at the corresponding moments shown in Figure 10 were drawn. The color meanings in Figure 10 are the same as those in Table 2, red arrows are used to indicate the temperature gradient directions, and red contours in the CFRTP composite indicate the boundaries of the melting area. In all moments, the y direction temperature gradients in the CFRTP composite were directed toward the aluminum alloy, while the direction of the x direction temperature gradients was related to the laser irradiation position. From 7.00 s to 7.14 s, the melting areas surrounded by the contours were basically the same as the areas of the CFRTP composite moving toward the aluminum alloy, indicating that the molten polymer drove the fibers to move toward the aluminum alloy under the large y direction temperature gradient. Meanwhile, since the second ply of the CFRTP composite was not melted, the first ply moved toward the aluminum alloy, causing delamination between it and the second ply. Once the melting of the polymer in the second ply occurred at 7.24 s, the molten polymer and fibers in the second ply moved toward the aluminum alloy, further proving that the temperature gradient in the melting area drove the CFRTP composite to move and the direction was related to the gradient direction. For the movement toward the laser scanning direction, although the temperature gradients from 7.00 s to 7.14 s were in the $x+$ direction, the observations showed that the polymer matrix drove the fibers to move in the $x-$ direction, whereas temperature gradients toward the $x-$ direction appeared in the observation area at 7.24 s. This was because the flow of the molten polymer in the CFRTP composite was influenced by the fibers; under the extrusion effect of the scanned area and the support effect of the unscanned area, the movement of the molten polymer and fibers in the x direction was affected by the movement which occurred away from the observation area.

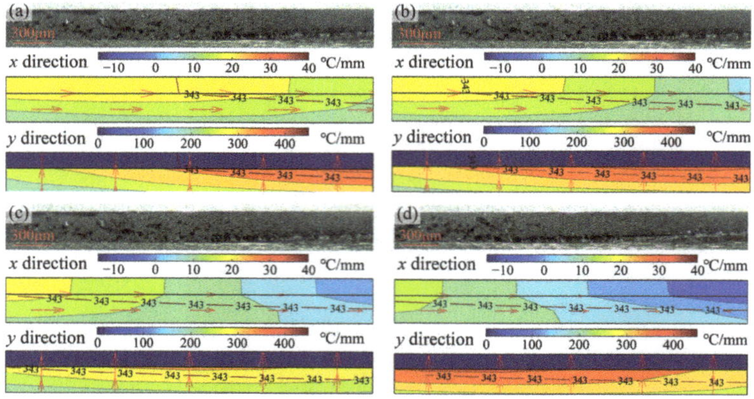

Figure 10. Interfacial bonding processes and x and y direction temperature gradient distributions at different moments when the polymer is unfilled: (**a**) 7.00 s, (**b**) 7.09 s, (**c**) 7.14 s, and (**d**) 7.24 s.

The interfacial bonding processes when the interface was filled with the polymer layer are shown in Figure 11, with a laser power of 600 W. The polymer began to melt at 4.85 s in the observation area, corresponding to the laser scanning direction from left to right. Due to the better melt fluidity of the pure polymer, the gas in the gap between the polymer and the aluminum alloy could not be vented in time, and bubbles were found in the molten polymer near the aluminum alloy in the observation results. As the laser scanned to the right, the molten polymer gradually filled the gap between the polymer and the aluminum alloy, accompanied by the movement of bubbles along the laser scanning direction, while there was no significant change in the position in the thickness direction. Compared with laser direct joining without filling the polymer layer, the polymer layer could rapidly infiltrate the aluminum alloy surface after melting. Once the molten polymer filled the gap at the material interface, the fixed aluminum alloy limited the movement towards the aluminum alloy. This resulted in the molten polymer moving mainly along the laser scanning direction, avoiding cracking and delamination in the CFRTP composite caused by the movement towards the aluminum alloy.

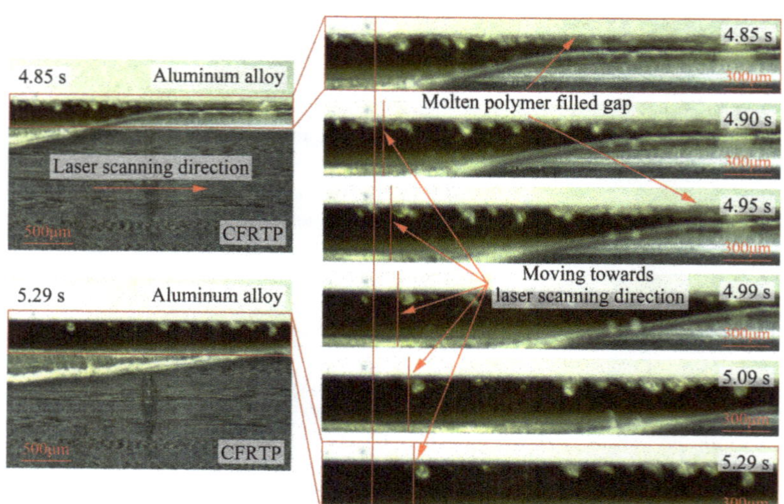

Figure 11. Interfacial bonding processes at different moments when the polymer was filled.

To reveal the motion of the molten polymer layer during the interfacial bonding process, the observations and temperature gradients in a single laser scan were analyzed, that is, the laser direct joining processes from 6.10 s to 7.90 are shown in Figure 12. The laser scanning direction was from right to left, and the laser irradiation position was on the sides of the aluminum alloy at 6.10 s and 7.90 s, while the laser irradiation position was on the sides of the observation area at 6.90 s and 7.09 s. From 6.10 s to 6.90 s, when the laser scanned the right side of the outer observation area, the bubble in the molten polymer appeared to move to the $x+$ direction, whereas the bubble position was unchanged when the laser scanned the observation area from 6.90 s to 7.09 s. Furthermore, the bubble moved to the $x-$ direction when the laser scanned the left side of the outer observation area from 7.09 s to 7.90 s. The x direction temperature gradients in the observation area followed the same trend as those of the bubble movement, that is, the x direction temperature gradient was positive when the bubble moved to the $x+$ direction, gradually decreased to zero when the bubble position was unchanged, and there was a negative x direction temperature gradient when the bubble moved to the $x-$direction at 7.09 s. The above results proved that the direction of the temperature gradient was the key to determine the flow of molten polymer, and the molten polymer flowed faster when the direction of the temperature gradient was the same as the laser scanning direction.

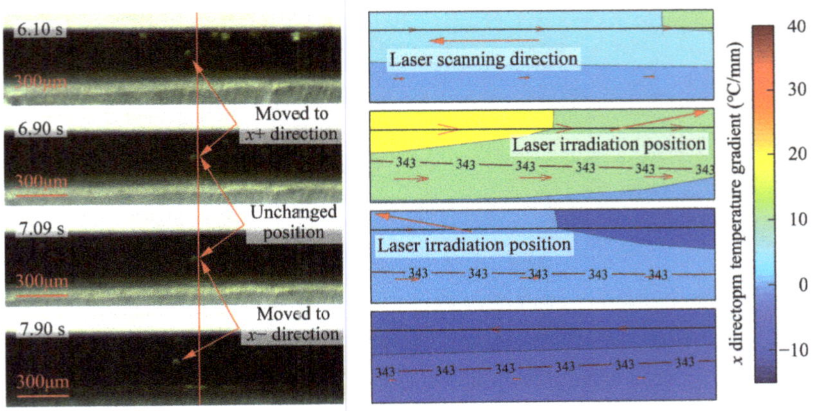

Figure 12. Bubble position change and x direction temperature gradient in single-laser scanning.

4. Conclusions

To investigate the interfacial bonding process between the CFRTP composite and aluminum alloy during laser direct joining, an online observation device for the interfacial bonding process was designed, and the temperature field simulation model for laser direct joining was established. Based on the observed and simulated results, the relationship between the temperature distribution and the melting and flow of the polymer was analyzed, as well as the effect of the filled polymer at the interface on the interfacial bonding process. The main conclusions are as follows:

(1) The melting and flow of the polymer matrix and filled polymer at the interface were observed, as well as the position change of the bubbles and fibers therein. As the laser power increased, the number of melted plies of the CFRTP composite increased when the interface was not filled with polymer. Once the interface was filled with polymer, the number of melted plies increased and then remained constant as the laser power increased, because bubbles caused by the thermal decomposition of the polymer layer at high laser powers hindered the heat transfer into the CFRTP composite.

(2) The number of melted plies in the CFRTP composite obtained with the simulation were the same as the observed results, when thermal decomposition of the polymer did not occur. Affected by the low thermal conductivity of the CFRTP composite and polymer,

the largest temperature gradient occurred at the contact area with the aluminum alloy. Regardless of whether the interface was filled with a polymer layer or not, the temperature gradient along the material thickness direction was larger than that along the laser scanning direction, which was about 400 °C/mm and 40 °C/mm, respectively.

(3) When the interface was not filled with polymer, the molten matrix together with the fibers moved to the aluminum alloy under the temperature gradient along the material thickness direction. This resulted in filling the gap between the CFRTP composite and the aluminum alloy, coupled with cracking between the matrix and fibers and delamination in the CFRTP composite. For the movement along the laser scanning direction, less movement of the molten matrix and fibers was observed because of the lower temperature gradient and the support effect of the unscanned area.

(4) When the interface was filled with polymer, the gap between the polymer layer and the aluminum alloy was filled rapidly once the polymer started to melt, after which the aluminum alloy limited the flow of the molten polymer towards the aluminum alloy, thus avoiding cracking and delamination. However, because of the better fluidity of the molten polymer, the gas in the interfacial gap tended to remain in the molten polymer. The molten polymer and bubbles moved in the laser scanning direction, and the motion was in the positive direction of the temperature gradient.

In the future, optimization and modification of the interface-filled polymer layer and development of the matched laser direct joining process are still needed. Those studies are expected to promote the flow of the molten polymer and bonding with aluminum alloy based on the suppression effect of the polymer layer on joining defects. The promoted interfacial bonding and reduced joining defects will contribute to the improved performance and application of laser direct joining between CFRTP composites and aluminum alloys.

Author Contributions: Conceptualization, Q.W., R.F. and F.W.; methodology, Q.W.; software, Q.W. and C.L.; validation, R.F. and J.L.; formal analysis, J.L.; investigation, Q.W. and C.L.; resources, R.F. and F.W.; data curation, R.F.; writing—original draft preparation, Q.W. and J.L.; writing—review and editing, F.W.; visualization, Q.W. and C.L.; supervision, F.W. and Z.J.; project administration, Z.J.; funding acquisition, R.F., F.W. and Z.J. All authors have read and agreed to the published version of the manuscript.

Funding: This research was funded by the National Natural Science Foundation of China, grant number 52090053, and the Science and Technology Innovation Foundation of Dalian, grant numbers 2021RD08, 2022JJ12GX027.

Institutional Review Board Statement: Not applicable.

Informed Consent Statement: Not applicable.

Data Availability Statement: Data can be obtained from the corresponding author upon reasonable request.

Conflicts of Interest: The authors declare no conflict of interest.

References

1. Siengchin, S. A review on lightweight materials for defence applications: Present and future developments. *Def. Technol.* **2023**, *24*, 1–17. [CrossRef]
2. Wu, J.W.; Posen, I.D.; MacLean, H.L. Trade-offs between vehicle fuel economy and performance: Evidence from heterogeneous firms in China. *Energy Policy* **2021**, *156*, 112445. [CrossRef]
3. Ahmad, F.; Al Awadh, M.; Noor, S. Optimum alternate material selection methodology for an aircraft skin. *Chin. J. Aeronaut.* **2023**, *36*, 476–488. [CrossRef]
4. Cao, J.F.; Bai, S.L.; Qin, W.Z.; Yan, Y. Research progress on preparation and properties of carbon fiber reinforced thermoplastic composites. *Acta Mater. Compos. Sin.* **2023**, *40*, 1229–1247.
5. Min, J.; Hu, J.; Sun, C.; Wan, H.L.; Liao, P.X.; Teng, H.; Lin, J.P. Fabrication processes of metal-fiber reinforced polymer hybrid components: A review. *Adv. Compos. Hybrid Mater.* **2022**, *5*, 651–678. [CrossRef]
6. Lin, Y.; Min, J.; Teng, H.; Lin, J.P.; Hu, J.H.; Xu, N.J. Flexural Performance of Steel–FRP Composites for Automotive Applications. *Automot. Innov.* **2020**, *3*, 280–295. [CrossRef]
7. Backe, S.; Balle, F. A novel short-time concept for fatigue life estimation of carbon (CFRP) and metal/carbon fiber reinforced polymer (MCFRP). *Int. J. Fatigue* **2018**, *116*, 317–322. [CrossRef]

8. Gardiner, G. A350 XWB update: Smart Manufacturing. Available online: https://www.compositesworld.com/articles/a350-xwb-update-smart-manufacturing (accessed on 7 October 2023).
9. Roux, M.; Eguémann, N.; Dransfeld, C.; Thiébaud, F.; Perreux, D. Thermoplastic carbon fibre-reinforced polymer recycling with electrodynamical fragmentation: From cradle to cradle. *J. Thermoplast. Compos. Mater.* **2017**, *30*, 381–403. [CrossRef]
10. Zhao, T.; Zhao, Q.Y.; Wu, W.W.; Xi, L.; Li, Y.; Wan, Z.S.; Villegas, I.F.; Benedictus, R. Enhancing weld attributes in ultrasonic spot welding of carbon fibre-reinforced thermoplastic composites: Effect of sonotrode configurations and process control. *Compos. Pt. B-Eng.* **2021**, *211*, 108648. [CrossRef]
11. Jin, Z.; Han, Z.; Chang, C.; Sun, S.Z.; Fu, H.Y. Review of methods for enhancing interlaminar mechanical properties of fiber-reinforced thermoplastic composites: Interfacial modification, nano-filling and forming technology. *Compos. Sci. Technol.* **2022**, *228*, 109660. [CrossRef]
12. Martín, I.; Fernández, K.; Cuenca, J.; Sánchez, C.; Anaya, S.; Élices, R. Design and manufacture of a reinforced fuselage structure through automatic laying-up and in-situ consolidation with co-consolidation of skin and stringers using thermoplastic composite materials. *Heliyon* **2023**, *9*, e12728. [CrossRef]
13. Behrens, B.A.; Brosius, A.; Hintze, W.; Ihlenfeldt, S.; Wulfsberg, J.P. Automated Assembly of Thermoplastic Fuselage Structures for Future Aircrafts. In *Production at the Leading Edge of Technology*; Brosius, A., Hintze, W., Ihlenfeldt, S., Wulfsberg, J.P., Eds.; Springer: Berlin/Heidelberg, Germany, 2021; pp. 467–475.
14. Omairey, S.L.; Sampethai, S.; Hans, L.; Worrall, C.; Lewis, S.; Negro, D.; Sattar, T.; Ferrera, E.; Blanco, E.; Wighton, J.; et al. Development of innovative automated solutions for the assembly of multifunctional thermoplastic composite fuselage. *Int. J. Adv. Manuf. Technol.* **2021**, *117*, 1721–1738. [CrossRef]
15. Gardiner, G. Automated Joining of Hybrid metal-Thermoplastic Composite Structures. Available online: https://www.compositesworld.com/articles/automated-joining-of-hybrid-metal-thermoplastic-composite-structures (accessed on 7 October 2023).
16. Murray, R.E.; Roadman, J.; Beach, R. Fusion joining of thermoplastic composite wind turbine blades: Lap-shear bond characterization. *Renew. Energy* **2019**, *140*, 501–512. [CrossRef]
17. Jiao, J.; Xu, J.; Jing, C.; Sheng, L.; Ru, H.; Xia, H. Laser welding process and strength enhancement of carbon fiber reinforced thermoplastic composites and metals dissimilar joint: A review. *Chin. J. Aeronaut.* **2023**, *in press*. [CrossRef]
18. Yang, H.Y.; Zhan, X.H.; Bu, H.C.; Ma, W.P.; Wang, F.Y. Study on bubble morphology at interface of laser direct joint between carbon fiber reinforced thermoplastic (CFRTP) and titanium alloy. *J. Polym. Eng.* **2020**, *40*, 868–875. [CrossRef]
19. Chen, J.; Bai, J.; Zu, S.; Lin, Z.; Yang, S. Effect of interfacial thermal history on bonding mechanism of laser assisted joining of QP980-CFRTP with adjustable flat-top rectangular laser beam. *Compos. Struct.* **2023**, *323*, 117488. [CrossRef]
20. Tan, C.W.; Liu, Y.F.; Liu, F.; Su, J.H.; Zhang, Z.Q.; Zhang, X.Y.; Song, K.J.; Chen, B.; Song, X.G. Effect of laser power on laser joining of carbon fiber reinforced plastic to AZ31B Mg alloy. *Opt. Laser Technol.* **2022**, *145*, 107449. [CrossRef]
21. Lambiase, F.; Genna, S.; Kant, R. Optimization of laser-assisted joining through an integrated experimental-simulation approach. *J. Adv. Manuf. Technol.* **2018**, *97*, 2655–2666. [CrossRef]
22. Lambiase, F.; Genna, S. Homogenization of temperature distribution at metal-polymer interface during Laser Direct Joining. *Opt. Laser Technol.* **2020**, *128*, 106226. [CrossRef]
23. Hussein, F.I.; Salloomi, K.N.; Akman, E.; Hajim, K.I.; Demir, A. Finite element thermal analysis for PMMA/st.st.304 laser direct joining. *Opt. Laser Technol.* **2017**, *87*, 64–71. [CrossRef]
24. Tan, C.; Su, J.; Zhu, B.; Li, X.; Wu, L.; Chen, B.; Song, X.; Feng, J. Effect of scanning speed on laser joining of carbon fiber reinforced PEEK to titanium alloy. *Opt. Laser Technol.* **2020**, *129*, 106273. [CrossRef]
25. Wang, Q.; Jia, Z.Y.; Zhang, B.Y.; Gao, D.L.; Ma, Y.; Liu, J.Y. Influence of processing parameters on joint shear performance in laser direct joining of CFRTP and aluminum alloy. *Mater. Des.* **2021**, *209*, 109996. [CrossRef]
26. Jiao, J.K.; Wang, Q.; Wang, F.Y.; Zan, S.P.; Zhang, W.W. Numerical and experimental investigation on joining CFRTP and stainless steel using fiber lasers. *J. Mater. Process. Technol.* **2017**, *240*, 362–369. [CrossRef]
27. Zhang, Z.; Tan, X.H.; Zhang, J.; Shan, J.G. Suppression of shrinkage porosity in laser-joining of CFRP and steel using a laser surface modification process "Surfi-Sculpt (R)". *Int. J. Adhes. Adhes.* **2018**, *85*, 184–192. [CrossRef]
28. Ai, Y.W.; Zheng, K.; Shin, Y.C.; Wu, B.X. Analysis of weld geometry and liquid flow in laser transmission welding between polyethylene terephthalate (PET) and Ti6Al4V based on numerical simulation. *Opt. Laser Technol.* **2018**, *103*, 99–108. [CrossRef]
29. Tao, W.; Su, X.; Chen, Y.B.; Tian, Z. Joint formation and fracture characteristics of laser welded CFRP/TC4 joints. *J. Manuf. Process.* **2019**, *45*, 1–8. [CrossRef]
30. Lambiase, F.; Genna, S. Experimental analysis of laser assisted joining of Al-Mg aluminium alloy with Polyetheretherketone (PEEK). *Int. J. Adhes. Adhes.* **2018**, *84*, 265–274. [CrossRef]
31. Meiabadi, M.S.S.M.; Kazerooni, A.; Moradi, M.; Torkamany, M.J. Laser assisted joining of St12 to polycarbonate: Experimental study and numerical simulation. *Optik* **2020**, *208*, 164151. [CrossRef]
32. Xue, Z.; Shen, J.; Hu, S. Influence of scanning speed and defocus distance on laser welded PA6GF30/SUS444 dissimilar lap joints. *Opt. Laser Technol.* **2022**, *153*, 108223. [CrossRef]
33. Jiao, J.K.; Jia, S.H.; Xu, Z.F.; Ye, Y.Y.; Sheng, L.Y.; Zhang, W.W. Laser direct joining of CFRTP and aluminium alloy with a hybrid surface pre-treating method. *Compos. Pt. B-Eng.* **2019**, *173*, 106911. [CrossRef]
34. Lambiase, F.; Genna, S.; Kant, R. A procedure for calibration and validation of FE modelling of laser-assisted metal to polymer direct joining. *Opt. Laser Technol.* **2018**, *98*, 363–372. [CrossRef]

35. Bu, H.; Zhan, X.; Yang, H.; Wang, F.; Wang, F.; Ma, W. Numerical simulation of thermal distribution and residual stress characteristic for laser wobble joining of CFRTP and Ti-6Al-4V alloy. *J. Manuf. Process.* **2022**, *79*, 562–575. [CrossRef]
36. Wang, Q.; Jiao, J.K.; Zan, S.P.; Zhang, W.W. Effect of Thermal Contact Conductanceon Temperature Field of CFRTP/Stainless Steel Laser Direct Joining. *Chin. J. Lasers* **2017**, *44*, 0402002. [CrossRef]

applied sciences

MDPI

Article

An Evaluation of the Tool Wear of Ceramic and Coated Carbide Inserts in Finishing Turning under the Influence of Age-Strengthening Gray Cast Iron

N. E. González-Sierra [1], Javier Flores Méndez [1,2], M. A. Meraz-Melo [2], Ana C. Piñón Reyes [2], German Ardul Munoz-Hernandez [1], Alfredo Morales-Sánchez [3], Mario Moreno Moreno [3] and Gustavo M. Minquiz [1,2,*]

1 Área de Ingeniería, Benemérita Universidad Autónoma de Puebla, Ciudad Universitaria, Blvd. Valsequillo y Esquina Av. San Claudio s/n, Col. San Manuel, Puebla C.P. 72570, México
2 Ingeniería Mecánica, Tecnológico Nacional de México/I.T. Puebla, Av. Tecnológico #420 Col. Maravillas, Puebla C.P. 72220, México
3 Departamento de Electrónica, Instituto Nacional de Astrofísica, Óptica y Electrónica, Luis Enrique Erro No.1, Sta. Ma. Tonantzintla, Puebla C.P. 72840, México
* Correspondence: gminquiz@yahoo.com

Abstract: Gray cast iron (GCI) is a common material in the automotive industry due to its mechanical characteristics, which change primarily for materials employed for the foundry and cooling rate of material. According to the workpiece, the material of the cutting tool and cutting parameters are analyzed to improve the machining and to increment the lifetime of the tools. In this research, the foundry and machining process of an automotive component using ceramic and coated carbide tools were the study case, and the effect that they have on the age strengthening of GCI on the tool wear of the cutting tools was studied. Both inserts have the capability to machine the material with a rough surface between 1.5 to 2.0 μm. The wear mechanism of inserts and the microstructure of GCI were characterized with microscopy techniques, atomic force microscope (AFM), and energy dispersive X-ray spectroscopy (EDS). The microstructure of the workpiece shows a casting with flake graphite morphology that is linked with the induction of microcracks in the material. The experimental analysis shows that the GCI with 12 days of aging has an increased tensile strength. This improves the tool life of ceramic and coated carbide tools. There is a 50% reduction in flank wear with inserts that are machined with the GCI within five days of aging, compared with the material within twelve days. The rake face and flank wear show that abrasive and adhesive wear are the main mechanisms of ceramic inserts due to the high cutting speed. Meanwhile, adhesive and oxidative wear in the flank were the predominant type of wear for coated carbide tools.

Keywords: gray cast iron; age strengthening; flank wear; rake face and microscopy

Citation: González-Sierra, N.E.; Flores Méndez, J.; Meraz-Melo, M.A.; Piñón Reyes, A.C.; Munoz- Hernandez, G.A.; Morales-Sánchez, A.; Moreno Moreno, M.; Minquiz, G.M. An Evaluation of the Tool Wear of Ceramic and Coated Carbide Inserts in Finishing Turning under the Influence of Age-Strengthening Gray Cast Iron. *Appl. Sci.* **2023**, *13*, 10248. https://doi.org/10.3390/app131810248

Academic Editors: Rui Manuel Leal, Gustavo H. S. F. L. Carvalho and Ivan Galvão

Received: 2 August 2023
Revised: 5 September 2023
Accepted: 8 September 2023
Published: 13 September 2023

1. Introduction

The foundry irons can be summarized in the following steps: (a) Metal melting: First, the melting process of various metallic materials and alloys is carried out to obtain gray iron. (b) Mold making: The disk mold is made by a process known as High-Density Vertical Molding. (c) Casting: This process consists of filling the mold cavity with molten liquid metal. After cooling, the metal solidifies and the sand is removed from the mold. At last, the cast material is transported to the machine shop [1]. Cast irons have a low melting point compared to steel, and these are carbon-rich iron-base alloys arranged near cementite or eutectic graphite. The GCI is usually machined in the as-cast condition, but the solidification process plays an important role. There are two variables that determine if the matrix could be austenite, ferrite, or a mixture of ferrite and perlite; those are the cooling rate and alloy composition [2]. Compacted Graphite Iron (CGI) presents a challenge in terms of machinability. An alternative to optimize the manufacturing process is the use of a metalworking fluid diluted between 4% and 9% and containing a 0.3 sulfur compound, with cutting lengths of 6.0 and 4.8 km [3]. The mechanical and microstructure performance of the GCI in a pearlitic state can be altered by the point of extraction foundry, even with the same composition. Additionally, the graphite content is determinant in mechanical behavior, and the fine structure of eutectic cells increases the fatigue limit as the phosphorous improves the tensile strength [4]. The hardness properties of the grey cast iron are determined by the cooling rate, which are related with the primary dendrite arm spacing (DAS), the secondary dendrite arm spacing (SDAS), and the thickness of the ferrite–cementite layer (l_e). While the cooling rate increases, the eutectoid thickness decreases. Eutectic cells are combined with interconnected graphite plates that surround austenite in the ferrite–cementite layers. An effect of the undercooling is the degree of ramification of graphite and the spacing [5]. The solid transformation of the GCI consists of releasing heat; this change of phase is divided into two points of interest that are described through a curve of the temperature as a function of the time. The first point occurs between 1140 to 1150 °C and 40 to 50 s; the collected data are related to the shrinkage behavior of the iron. The undercooling at the second point and graphite growth beyond are determined by the onset of the eutectic solidification of graphite and austenite. The mentioned phase occurs between 1130 to 1140 °C and 50 to 100 s [6]. The gray cast iron properties and microstructure were analyzed in a foundry process under different cooling methods. The study tested four approaches: as-cast, furnace-cooled, air-cooled, and green sand-cooled. The results showed the variation of cementite Fe_3C with 24.96, 11.64, 6.62, and 6.35 percent respectively, and the consequence is the differences in the hardness Rockwell scale of 95.8, 85.2, 94.7, and 97.0 HRB [7]. As the foundry process progresses, the crystal structure begins to re-form on cooling at the melting point, especially in solidifying when energy is given out in the form of the latent heat of fusion. Solidification depends on the rate at which heat can be removed. While the solidification is in a slow rate, the first nuclei increase in size as more eventually has transformed into large grains. Initially, the grain boundaries represented the points of growth outward from the nuclei. The number of clusters increases when cooling is rapid, and each cluster grows rapidly as a result. As more grains are formed in the solid, the metal is finer [8]. A constant temperature gradient is shown in Figure 1, which represents the solidification temperature increase as a function of the solidification rate for the Fe-4.28C eutectic composition. Possible solid constituents include g/graphite eutectic, g/cementite eutectic, and g/austenite dendrites. The dotted line indicates the phase with the highest interface growth temperature. The g/Graphite eutectic forms with a low solidification rate, followed by austenite dendrites, whereas the g/cementite eutectic forms with a high solidification rate [2].

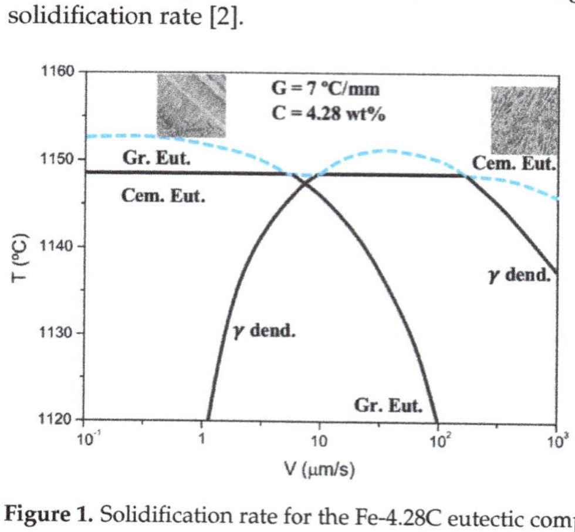

Figure 1. Solidification rate for the Fe-4.28C eutectic composition [2].

Age strengthening is the final phase of the foundry process in gray cast iron, where the mechanical properties are defined. Tool wear and vibration are the principal issues during the machining process; the common solution had been stocking parts for a few days to reach metallurgical stability. A brake disk of gray cast iron was analyzed for hardness during aging, and the data collected showed that the metallurgical stability was reached at seven days. The investigations exhibited the importance of industrial conditions before machining the casting [9,10]. An investigation has been conducted into the influence of age on the machining of a component of an automobile with a polycrystalline cubic boron nitride (PCBN) cutting tool. The test indicated that the tendency to improve the tool life is not under control during the first 100 h. After this time, there is a stable trend [11]. The age-strengthening effect in gray cast iron was analyzed, in terms of ultimate tensile strength, and the data collected showed a variation between 3.3% and 13% [12]. The age strengthening appears as nitride precipitation in ferrite. It appears that elements that interact with N, such as Mn, disturb the strength of the aging process, and the gray cast iron's machinability improves along with age-strengthening, particularly with free ferrite present [13]. Some foundry companies have established that five-day to room temperature aging provides a good balance between production and shipping and can require floor space [14].

A review of the literature found that the nature and temporal behavior of age strengthening are supported in two relevant aspects [13]. First, during the melting process, free nitrogen into iron-based alloys is required for the aging effect. Because this influence of the nitrogen could be related with the precipitate and diffusivity in the microstructure, which is validated to the fact that nitrogen is better diffused when the casting is cooling at room temperature [15]. In addition, the weight percent of iron–nitride (Fe_4N) is linearly correlated with the changes of the ultimate tensile strength (UTS) from age strengthening, which implies that Fe_4N significantly promotes the precipitate [16]. Second, the tests realized the effects of the aging temperature, which have demonstrated that it is possible to increase the rate of age strengthening when immediately after the casting, a temperature of 182 °C or 285 °C is applied to the GCI. For the last case (285 °C), the aging is accelerated through reaching a peak in tensile strength in approximately five hours [9,17]. However, after five hours, if the process of heating continues, over-aging occurs that decreases the tensile strength [18]. In contrast, the same effects can take any amount of days or a few weeks for the aging of GCI at room temperature. In this work, the workpiece was cooled at room temperature, and accordingly, the aging was a slow process. Additionally, it can be deduced that the influence of nitrogen exists in casting because the aging workpiece improves the machining.

Based on previous research, casting material properties are defined considering the cooling rate or the influence of the content metals in the foundry process. Even more, age strengthening of the cast iron defines the casting mechanical properties. The present study aims to evaluate the tool wear mechanisms after five and twelve days from casting. As a case study, finishing turning operations with automotive material were tested using two cutting tool materials. A tool wear microscopy analysis was performed to examine the difference between ceramic and coated carbide inserts. This comparative considers the variation in age strengthening that occurs with the GCI.

2. Review of Wear Mechanism in a Grey Cast Iron Machining Process

Different cutting tools have been used in the machining of gray cast iron. The material from which the tool is made is a fundamental characteristic because it shows the effects of the interactions between the cutting tool and the workpiece [19]. Tool life optimization is a concerning discussion in a machining process, the change of cutting parameters affects the performance of others, and there is a complex relationship. Consequently, the cutting tool characteristics must have superior toughness, hot hardness, thermal shock resistance, oxidation resistance, and affinity with the material to be machined [20]. It is possible to machine with ceramics and PCBN tools with good productivity or quality surface, and the cutting speed parameter can be between 500 and 2000 m/min [21]. The machinability of the PCBN chamfered tool in the dry turning of gray cast iron has been investigated. To evaluate the tool wear mechanism, the cutting parameters were fixed as V_C = 204 m/min, a_P = 0.4 mm, and f_n = 0.2 mm/r for the cutting speed, the depth of cut, and the feed rate, respectively. The main wear mechanisms found were material adhesion and microchipping due to the negative rake angle on the tool chamfer. The embedded material on the flank face suffered severe friction during the cutting process, causing some of the tool material to drop off, and this will chip, forming micro notches. An energy dispersive X-ray spectroscopy (EDX) analysis of the adhered material on the tool confirmed the presence of elements such as Fe, Si, and Mn, which correspond to the composition of gray cast iron [22]. The cutting properties of boron nitride dispersed cemented carbide (BNDCC) inserts that are a hybrid between cemented carbides, and cubic boron nitride (CBN) cutting tools have been analyzed. The tool wear was evaluated during grooving spheroidal cast iron by varying cutting speeds, such as $V_C \leq 317$ m/min for coated and $V_C \leq 176$ m/min for uncoated BNDCCs. The results show that adhesion wear predominates in the different types of BNDCC inserts. The build-up edge (BUE) is significantly major in uncoated BNDCC inserts, which is attributed mainly to the high friction in the tool flank-workpiece interface. In comparison, the BUE formation for coated (with TiN and TiAlN) BNDCC inserts is associated with the characteristics of the coating, which increases the coefficient of friction between the chip and gray iron cast [23,24]. The PCBN cutting tools have been used for the high-speed machining of EN-GJL-250 gray cast iron because they increment the tool life and material removal rate. A paper showed the tool wear mechanisms with PCBN of different compositions and geometry. The cutting data of the tests are: V_C = 1300 m/min, f_n = 0.5 mm/r, and a_P = 0.7 mm. Under these conditions, the results of scanning electron microscope (SEM) observations showed an adhesion layer strongly attached in the tools, which were cleaned by etching in an ultrasonic bath to be certain of the measure of the flank wear. This built-up layer can act as a medium of tool protection during turning [25]. A cemented carbide insert WC–5TiC–0.5VC-8Co (WTVC8) was tested on a machining gray cast iron (HT250). In this case, the wear morphology and EDS analysis conclude that adhesive wear was the principal wear mechanism on the flake face in the cutting tool [26]. A carbide tool SNMA 12408 chemical vapor deposition (CVD), coated with titanium carbonitride, aluminum oxide, and titanium nitride, was tested on machining gray cast iron. For understanding the sulfur influence, the cutting parameters configured were V_C = 100, 150, and 200 m/min, f_n = 0.257 mm/r, and a_P = 2 mm. The wear mechanisms were the adhesion and chips on the tool face with casting machining with a 0.065% sulfur content; furthermore, the tool life was reduced to 24, 32, and 38%, respectively, with each cutting speed tested [27]. For machining CGI, SNMA 120408 KR 3205 and 3210 inserts were tested. The cutting tools have a multilayer coated carbide of Ti(CN) and Al_2O_3 as exterior coating and an additional thin TiN coating. The parameters set were V_c = 250 m/min, f_n = 0.2 mm/rev, and a_P = 2 mm, and the results of the configuration were a few chippings and a more extensive flank and crater wear [28]. The relation between the flank-wear decreasing when cutting speed is above 900 m/min was studied, and the behavior was explained with a CBN cutting tool. The degradation happens because the oxidized Fe reacts with CBN and from B_2O_3 [29]. The insert NP-TNGA 160412GS3 MB710 with a cover layer of TiC and Al_2O_3 is an alternative, analyzed in the GCI machining. This tool was experimented with in orthogonal arrays, and the parameters are V_C = 750 to 1050 m/min, f_n = 0.05 to 0.15 mm/rev, and a_P = 0.1 to 0.2 mm. The combined analysis of thermal, SEM, and EDX has shown that tool life increases when the cutting speed does not exceed 850 m/min when using CBN inserts. This is explained by the phase transformation of the workpiece material from the pearlite to the austenite phase. In addition, chemical degradation and tool life decreased when the insert reached 1000 °C [30]. Based on previous research, grey cast iron has been machined with ceramic or coated carbide tools with a good surface finish.

3. Procedures for Experiments

The material used in this research is a cast iron metal workpiece that will eventually become a brake disk. The sequence followed by the industry process to produce the brake disk is shown in Figure 2a. Two options are available: 5 or 12 days of age-strengthening casting. The machinery worker decides between ceramic and tungsten carbide tools. Due to the benefits of the reduced cycle time, the manufacturer selected cutting parameters according to the recommendations of the toolmaker while maintaining a maximum cutting speed and feed per tooth. The brake manufacturer utilized a CNC (Puma V400, Doosan, Seoul, South Korea). The spindle rotates at 3000 rpm, and the maximum feed speed is 2000 m/min. Figure 2b summarizes the experimental process carried out in this work. The first step is to analyze the workpiece using an optical microscope and EDS mapping to study its morphology and the distribution of its elements. Second, the cutting tool was characterized with microscopy techniques for identifying the type of wear and the flank wear. Additionally, the adhered material on the inserts was inspected using a scanning electron microscope (SEM) and an energy dispersive X-ray spectrometer (EDX). To identify the damage zone in the inserts, a microscope (B-Serie, Optika, Ponteranica, Italy) was used to measure the flank wear at a 10 X fair value to the 100 X real size. Cutting tool samples were placed on a base with their flanks facing the instrument lamp, and the corner radius was measured starting from the edge to the damage inside. The details on the flank wear and rake face were analyzed using a SEM (SU3500, Hitachi, Tokio, Japan) with 10 kV and a magnification of 39.5 × 170 mm. The flank wear was measured and compared according to the specifications of ISO 3685 [31], which describes that the tool life admissible is 0.3 mm in carbide and ceramic indexable tools. Surface roughness was acquired through the measurement with the atomic force microscope AFM (EasyScan, Nanosurf, Liestal, Switzerland). The instrument is set up on a non-contact mode, and the probe model is configured with a tapping mode AFM probe and a long cantilever. Then, an oscillating AFM probe senses long-distance van der Waals forces. The EasyScand DFM version 2.3 software generates the images.

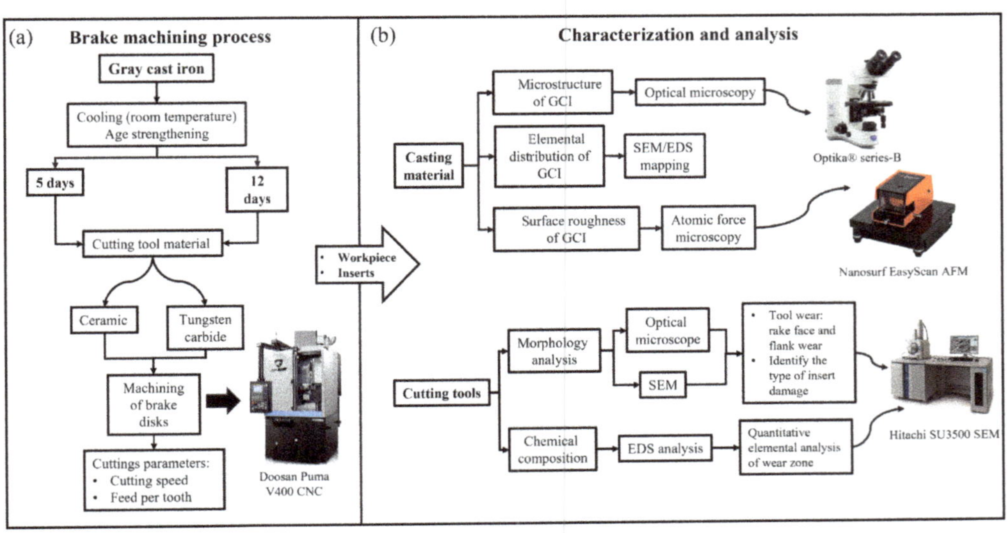

Figure 2. Methodology for brake disk and tool wear analysis: (**a**) brake machining process; (**b**) characterization and analysis.

Brake Machining Process Details from the Manufacturer

Table 1 shows that the CNMG is an insert of a carbide substrate with chemical-coated CVD. This coating consists of a hard top layer combined with Al_2O_3 and carbon-rich TiCN base layers as the main features of this tool [32]. The SCGN insert (Ceramic,CeramTec,Plochingen, Germany) is based on standard silicon nitride ceramics grades SL 500 [33].

Table 1. Tool data.

Indexable insert cutting tool:	- CNMG120412TSF- T515 Coated Carbide Grade (CVD), (Coated Carbide, Tungaloy, Fukushima, Japan)—ISO Range—(K15) Effective cutting edge: 4, Chip breaker: 13-degree Rhombic insert, 80-degree with hole, Corner radius: 1.2 mm, Negative relief angle. Cost: 15 USD - SCGN090412E-F Grade: SL 500, (Ceramic,CeramTec,Plochingen, Germany). Clearance angle major: 7-degree, Effective cutting edge: 4 Corner radius: 1.2 mm. Cost: 21 USD

An automotive brake disk is the study case, and the working surface of the disk is the study section. Figure 3a shows the shape and diameter of the workpiece, and the cut line sections indicate the solid material in the brake. Figure 3b indicates with the arrow to the remotion material area, which has an area of 10,162.4 mm^2, considering that the parameter of the depth of the cut is 1 mm because it is finishing turning.

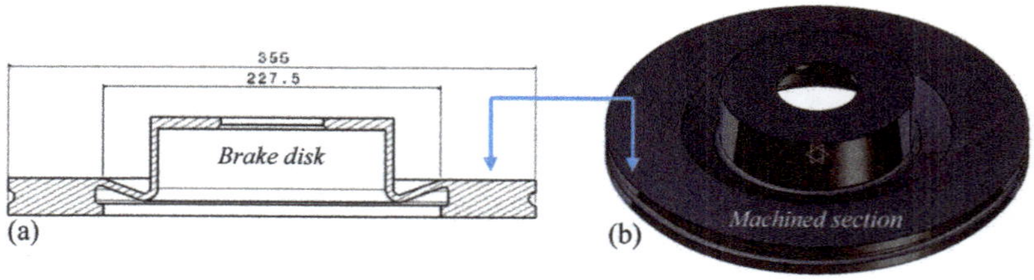

Figure 3. (**a**) Brake disk geometrical features; (**b**) area machined.

The choice of cutting tool type includes the shape and geometry of the cutting edge, which affects vibration and tool life. Additionally, it is considered the insert material or the combination of materials that constitute the cutter's tool edge. Then, the supplier proposes the cutting parameters in a range of values for cutting speed, depth, and feed per tooth of the insert according to its characteristics. The data is where the insert has the best compromise to remove the material, and it is necessary to calculate the spindle speed and feed rate, which determine the cycle time and quality of the surface [20,34]. Due to the aging of the GCI conditions, the brake manufacturer applied a practical solution to reduce chatter and maintain a rough surface between 1.5 and 2.0 µm, resulting in the use of two inserts, SCGN and CNMG, even though the inserts are made of different materials. According to the tool manufacturer, the SCGN insert works with cutting speeds up to 1200 m/min. In the ceramic cutting tool, the brake maker chose a parameter of 1115 m/min, which is a high value (Table 2). For both cutting tools, the constant parameter was the depth of cut at 0.1 mm. On the other hand, the coated carbide insert CNMG was configured with 650 rpm and 724 m/min, while the tool suppliers recommended cutting conditions between 150 to 700 m/min. The insert machined 40 workpieces with a cycle time of 0.18 s. The two inserts were configured in one pass because it is a finished operation.

Table 2. Configuration of cutting parameters.

Tool	Insert	rpm	v_c (m/min)	f_z (mm)
1	CNMG120412TSF-T515	650	724	0.1
2	SCGN090412E-F	1000	1115	0.1

4. Casting Material Analysis

For examining the microstructure of the GCI, the metallographic preparation of a specimen of GCI was implemented. First, a piece of the brake disc was cut with dimensions of 10 × 10 mm. The sample was roughed and polished with wet sandpaper of a grit of 240, 320, 400, 600, 1000, and 2000. Finally, the fine polish with a cloth and alumina powder (0.5 µm particles) was realized, although the grinding scratches were difficult to remove in this step (see Figure 4). Etching reagents were not used due to it being possible to see the morphology of the microstructure clearly with the optic microscope after the final polished. The austenite matrix has transformed to ferrite (light), and the presence of the flake graphite can be seen in Figure 4.

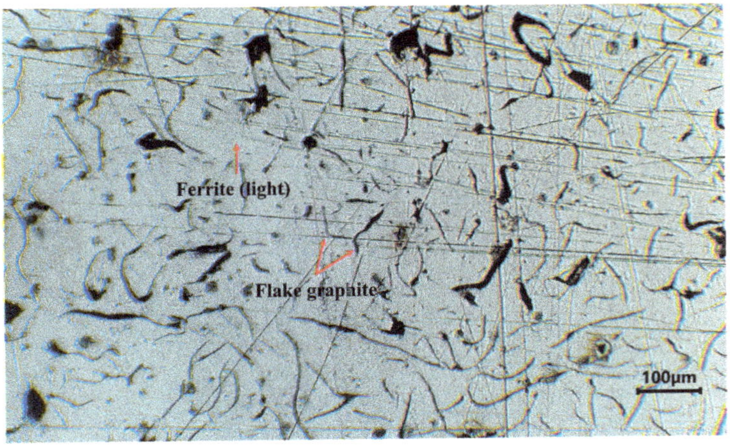

Figure 4. Gray cast iron showing a flake graphite morphology, unetched.

The foundry company provided data on the chemical composition and mechanical properties of gray cast iron that are shown in Table 3. Both tensile strength and Brinell hardness of 120 to 288 h of age strengthening of GCI at room temperature exhibited a clear increase. The tensile strength increases 3 MPa; meanwhile the hardness changes to 183 HB to 184.2 HB after 12 days of aging. These values show that a perceptible modification of the tensile strength of GCI after a week of aging occurs. In contrast, other results indicate an insignificant variation [18]. The values of tensile strength and of hardness support, from the technical point of view, the effects of aging on the machinability of GCI due to being correlated.

Table 3. Gray cast iron characterization.

Workpiece Material	Grey Cast Iron										
	5 days aging						12 days aging				
Hardness	183 HB						184.2 HB				
Tensile strength (N/mm^2)	177						180				
Chemical composition (at.%)	C 3.8	Si 2.44	P 0.01	Mn 0.65	S 0.11	Mo 0.01	Cr 0.23	Cu 0.05	Ni 0.02	Sn 0.09	Ti 001

To determine the composition of grey cast iron, a piece section of a brake disc was cut. The sample surfaces were cleaned with alcohol and measured with an electron microscope. Figure 5, in the red square on the left, shows the section analyzed with an approximation of 200 mm, belonging to a machined surface. Iron constitutes 77.43% of the mass, followed by oxygen at 11.75% and carbon at 6.95%, visible in the right section of Figure 5.

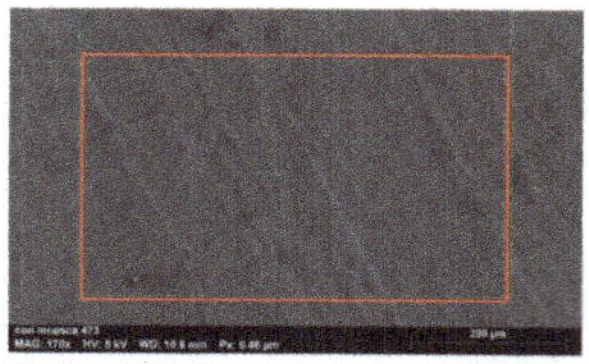

Element	At. No.	Netto	Mass [%]	Mass Norm. [%]
Carbon	6	27,494	6.95	6.95
Nitrogen	7	2578	0.83	0.83
Oxygen	8	55,401	11.75	11.75
Silicon	14	7575	2.06	2.06
Chromium	24	1627	0.98	0.98
Iron	26	188,238	77.43	77.43
		Sum	100.00	100.00

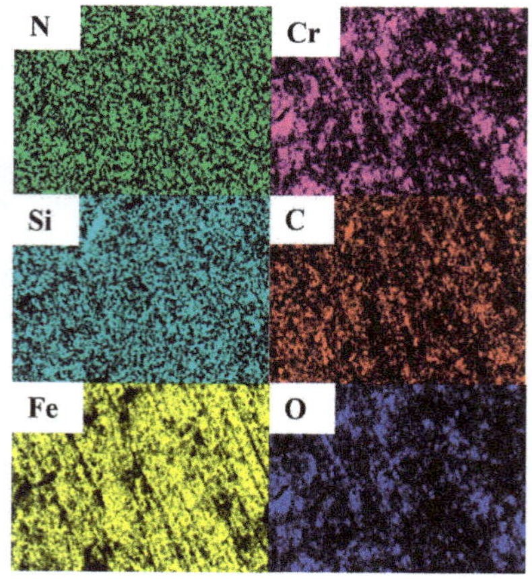

Figure 5. EDS Map of GCI.

5. Cutting Tool Characterization Results

After machining 40 workpieces, the morphology wear was revised with an optic microscope. The results of the microscopy analysis were categorized based on the cutting tool material and the aging time of the workpiece. The Figures 6–8 were labeled to identify the insert type and age of the workpiece. The acronyms SCGN and CNMG refer to the ceramic and coated carbide insert, respectively, and the symbol five or 12 at the end of the acronym represents the age strengthening of the cast iron, and the letter F or R represents the flank wear or rake face. The research tested four inserts, and two of the cutting tools were ceramic, which is on the top side of Figure 6, and flank wear displays the contrast between machining materials with five or 12 days since they left the foundry. The notch wear and crater were the representative damage in the tool that machined the materials within five days of aging. In contrast, the cutting tool used to machinery with the material with 12 days of aging has a normal flank wear and material adhesion.

Furthermore, the performance of the cover carbide tool is at the bottom in Figure 6. The chipping and adhesion material was on the flank wear and rake face, considering the insert that machined the material with five days of age straightening. As a result of the machining of the material 12 days after the casting process, normal flank wear with adhesion combined with plastic deformation was observed.

In Figure 7, an isometric view of the cutting tools is presented. Figure 7-SCGN5F shows that the tool has deep grooves on the flank due to the abrasion mechanism; a similar result of the wear type was observed in previous research for ceramic and PCBN cutting tools [35]. In comparison with this work, SCGN5F exhibits a severe abrasion that could be a consequence of the cutting speed of the tool and the aging time of the workpiece. The cutting speed is the same for SCGN5F and SCGN12F. Therefore, the wear type is associated only with the piece of GCI according to the aging time [12]. Consequently, for the machining of a workpiece of five days of aging time, the SCGN5 exhibited a flank wear of major abrasion compared with SCGN12F. Nevertheless, the ceramic cutting tool is more resistant to abrasion than the carbide coated insert [19]; the workpiece becomes more abrasive in relation to the aging time, which causes the hard particles to appear in the material. The hard particles can be detached of the workpiece when the tools move along the contact zone, causing abrasion in the cutting direction.

The most desirable tool failure is a flank wear uniform and smooth, as shown in the Figure 7-SCGN12F. This type of wear results from abrasion and exhibits a predictable behavior that changes progressively with time. In addition, the machining of the GCI piece of 12 days of aging time improved. Unlike SCGN5F, the worn flank of the SCGN12F does not show a serious crater in the corner and its flank wear is ~0.150 mm, which is approximately three times more mincr than the flank wear of SCGN5F, which represents a value permissible for ceramic cutting tools.

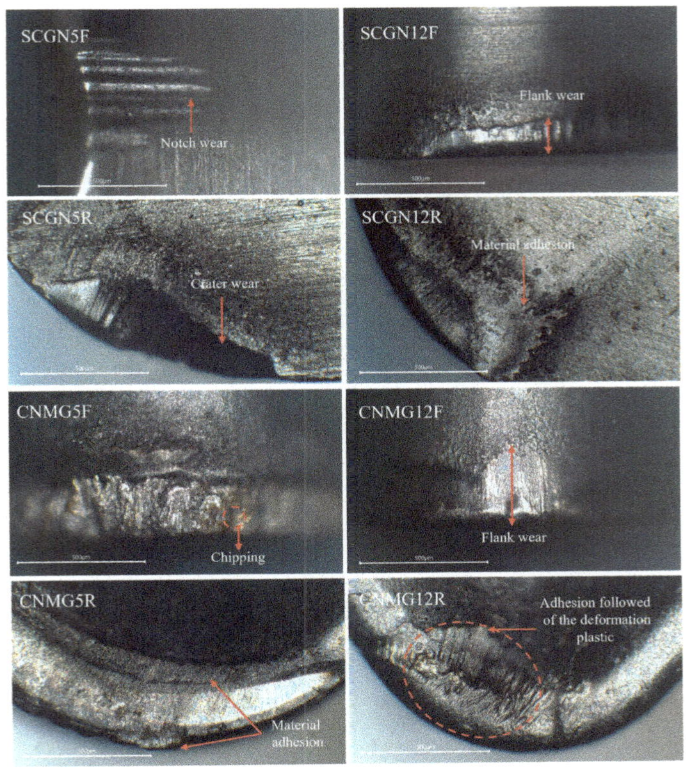

Figure 6. Tool wear analyzed by optical microscope.

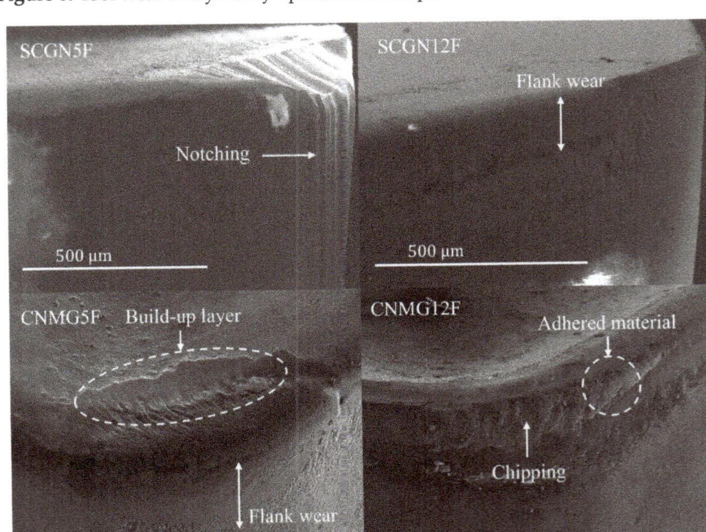

Figure 7. SEM micrograph of cutting tool flank wear.

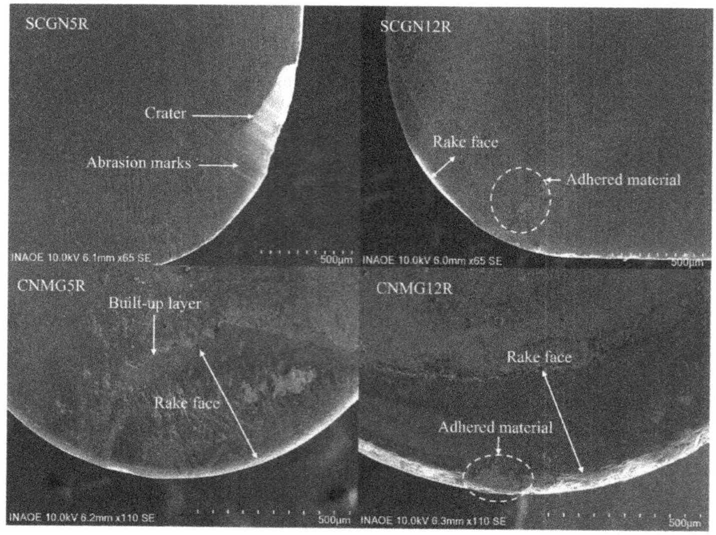

Figure 8. SEM micrograph of cutting tool rake face.

The morphology of the rake face and flank of the coated carbide tools is shown in the Figure 7 (CNMG5F and CNMG12F). The CNMG insert consists of a coating of a double layer (Al_2O_3 and TiCN) that wears progressively (see Figure 6-CNMG5R) due to the high cutting speed and the characteristics of the GCI, which is more abrasive and adhesive at five days of aging. In contrast to ceramic tools, serious adhesive wear (BUE formation) on the rake face and a minor amount of material adhered on the flank of the CNMG5F were observed.

The morphology of the tools examined for the optic microscope and SEM indicate that the main wear mechanism in the flank and rake face of the SCGN5F and SCGN12F inserts is abrasive wear. In contrast, the principal wear mode of the CNMG5F and CNMG12F tools was adhesion and oxidative wear on the flank and the rake face.

The behavior of the inserts that machined the material with an aging time of five days showed that the average flank wear for ceramic and coated carbide tools was 0.56 mm and 0.38 mm, respectively. For SCGN5F inserts, this value exceeds the general recommendations for the flank wear limit VBB, which is 0.3 mm according to ISO 3685:1993 [31]. The cutting tool used to machine the casting material within five days resulted in the highest flank wear compared with the 12 days material. The CGI age strengthening improves the machinability process once the casting is at an ambient cool temperature for more than 100 h. The UTS of the casting makes a better machining performance with extensive age strengthening [36].

Figure 8 displays the rake face images with a scale of 500 mm. Figure 8-SCGN5R displays the ceramic insert on the rake face zone. The gray cast iron with aging for five days was machined. The damage that appears with the crater wear on the image is caused by a chemical reaction between the workpiece material and the cutting tool, magnified by a 1115 m/min cutting speed. Figure 8-SCGN12R shows the rake face and adhering material along the cutting zone, and the rake face has a uniform flank wear with no vibration. Figure 8-CNMG5R presents a rake face, and it has layers of material with accumulation along the surface but is not a uniform distribution of materials. The rake face covers 0.45 mm from the edge to the inside of the tool. Figure 8-CNMG12R shows the conditions of the cutting tool that removed the material after age-strengthening for twelve days. In addition to the adhesion material along the working edge, there is also microchipping, which is the breaking of the cutting edge, caused by the high speed of the cutting tool. However, the rake face is uniform.

EDS Analysis

The chemical composition of the adhered material on the rake face and flank of the CNMG and SCGN inserts was checked through an EDS analysis. The concentration of each element was measured by the intensities of peaks, which was expressed in an atomic percentage, as shown in Table 4. The ceramic tools showed abrasion followed by adhesion as the main wear mechanisms according to the microscopy analysis. The EDS analysis on the surface of the SCGN12F insert shows a significant amount of Fe, Mn, Si, and C that confirmed the presence of the adhered material. However, the atomic percentage of the chemical elements found on wear zones of SCGN tools were not decisive for validating the phase composition of oxides as FeO or SiO_2. The ceramic tool cut at a speed of 1115 m/min, resulting in a temperature increase of up to 1000 °C due to the constant and rapid interaction between the tool and the chip. Under these conditions, a chemical oxidation reaction can occur between the Fe, the ceramic insert, and the oxidation during the cutting process. This too could explain the generation of notch wear observed in the SCGN tools [29,30,37].

Table 4. EDS analysis of rake face and flank wear.

SCGN5R		SCGN5F		SCGN12R		SCGN12F		CNMG5R		CNMG12R	
Element	**at. (%)**	**Element**	**at. (%)**	**Element**	**at. (%)**	**Element**	**at. (%)**	**Element**	**at. (%)**	**Element**	**at. (%)**
C	41.61	C	75.70	C	54.97	C	17.75	C	16.04	C	5.14
N	20.50	O	22.46	N	3.57	O	10.33	O	52.15	O	69.83
O	32.40	Al	0.11	O	39.28	Mg	0.91	Mg	0.10	Mg	3.28
Mg	1.78	Si	1.52	Mg	1.68	Al	5.46	Al	27.40	Al	2.28
Al	0.21	Zn	0.21	Al	0.02	Si	32.46	Si	0.63	Si	2.48
Si	2.75			Si	0.39	Mn	15.45	Fe	3.65	Fe	15.68
Zn	0.01			Mo	0.07	Fe	15.56	Zn	0.01	Mo	1.31
Mo	0.14					Zn	0.20	Mo	0.03		
						Mo	1.90				

The oxygen was found in all wear regions, mainly a large percentage on the flank of CNMG5 and CNMG12 inserts. The oxygen was formed during oxidation in air at a high temperature in the cutting zone, which may contribute to tool wear. For the case of coated-carbide tools, the inspection on the rake face of CNMG5R and CNMG12R indicated a higher concentration of O, Al, and Fe. In the case of the CNMG5R insert, the EDS analysis enabled the phase composition that was identified as alumina (Al_2O_3). The atomic percentage of Al and O was similar to the theoretical amount calculated for each element of the chemical form of alumina, considering the balance with the other elements that are shown in Table 4. Similarly, on the rake face of CNMG12, Fe and O were found with a quantity of 15.68 at.% and 69.83 at.%, respectively. These quantities suggest the presence of the phase of FeO as a wear product, which confirmed our previous analysis that the adhesion on the rake face is the main wear mechanism. Compared with other work, the aluminum and iron oxidation too have been observed [38].

In the Figure 9, a zinc peak can be seen in the EDS spectrum of the SCGN insert, which is unexpected because the certainty of amounts below ~0.1 at.% is more complicated with this method. Additionally, this could be associated with the accidental contaminants or with that identified of a chemical element wrongly due to the content of the elements obtained through an automatic quantification EDS.

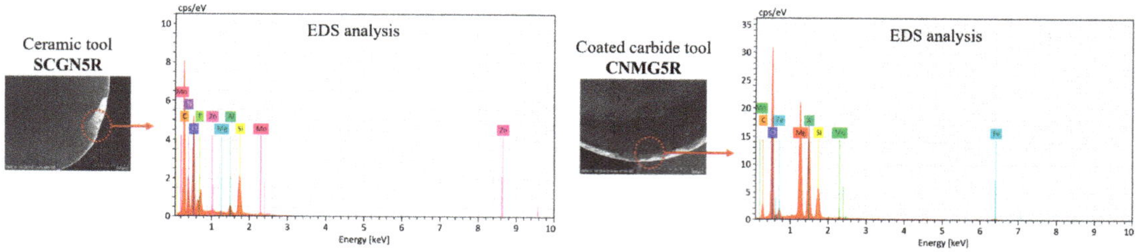

Figure 9. EDS spectrum of rake face of ceramic and coated carbide cutting tool.

6. Morphology of the Surfaces Machined

The surface roughness of the workpiece is one of the variables that determines its functionality. Nevertheless, the profile shape provides data relating to the contact between the cutting tool and the workpiece, and the topography contains the characteristics to make this analysis. The surface roughness analysis was carried out using a piece of the brake disk, but the studied area was 8 mm × 8 mm. From the AFM data, a rougher surface of the brake disk can be seen with the finishing turning, with a topography of peaks and valleys (Figure 10). Two samples were analyzed with 12 days of aging because this machined material showed better tool-wear behavior. The topography of the material machined with the SCGN tool has a slight elevation, and the few waver peaks and valleys are the constant topography. Also, the mean surface roughness is Sa 0.0327 µm, and RMS roughness is 0.052 µm (Figure 10a). The data collected display more variation in waver peaks and valleys with the material machined with the CNMG insert, and the results of the Sa and RMS roughness are 0.0467 and 0.06 µm for each one (Figure 10b).

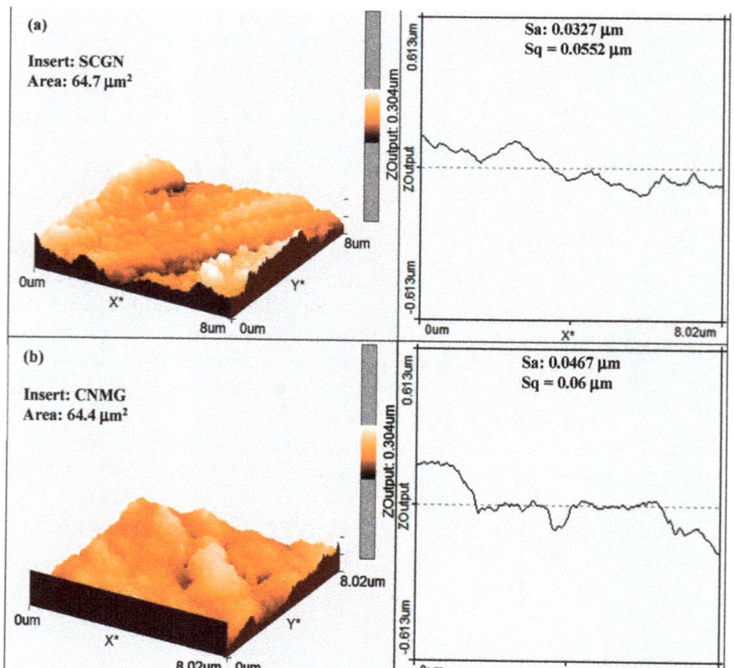

Figure 10. Surface topography of the workpiece machined with 12 days of aging: (**a**) ceramic insert, (**b**) coated carbide insert.

The surface analysis shows close values in Sa or Sq between the use of ceramic or coated carbide tools in the machining of GCI. However, in the topography portrait difference behavior, as shown in Figure 10, the uniformity in the wave's peaks and valleys is better with the SCGN tool, but these results are with the tool with fewer built-up layers. Otherwise, the CNMG does not have uniform topography, and the surface has a dispersity behavior. Even with this comparison, the geometrical tolerance is 1.5 µm, which means that both samples analyzed are under the quality surface established by the automotive companies.

It costs five dollars more to use the SCGN than the CNMG cutting tool. Furthermore, it is related to cycle time or quality surface, and the manufacturer has to trade-off between costs, time, and the quality process. The use of two different inserts, despite their dissimilarities, is a realistic solution to roughness surface issues in the brake disks and short tool life.

7. Conclusions

The present research examined the machining of GCI with an aging time of five or twelve days. The study case used two different cutting tools: CNMG and SCGN inserts. Both inserts have the characteristics to machine the material, but the finished workpiece must have a rough surface between 1.5 to 2.0 µm. An analysis of tool wear was carried out based on this information. The methodology analysis includes the rake face and tool wear measurement. In addition, it includes the characterization of the GCI and the topography surface of a brake disk.

The data collected from these two inserts were determinant to establish:

- The flank and rake face of the ceramic inserts showed abrasion (notching) as the main wear mechanism, which was more evident in relation to the age–strengthening variation of GCI. In contrast, the coated carbide inserts exhibited a predominantly adhesive wear and oxidation on the rake face, compared to the other wear mechanism. The oxidation resulted in wear products such as aluminum oxide and iron oxide, which were deduced from the quantitative EDS analysis on the rake face of CNMG tools.
- Five and twelve days of aging in GCI affected 50% of the tool life of a coated carbide insert. Based on this, 1.2 mm was the highest flank wear measured by the SEM analysis.
- From the data collected by AFM, it is possible to correlate the effect of the adhesive material on the cutting tool with the final surface. The machining of GCI with a coated carbide insert resulted in a slightly nonuniform topography in contrast to the ceramic cutting tool.
- The authors suggest that twelve days of age strengthening would allow the use of a carbide insert with uniform wear damage when time is not an issue in the machining shop. Manufacturers should evaluate time and cost before selecting cutting tool materials.

This investigation helps to further understand the tool wear mechanisms of the two inserts that were analyzed when completing the machining of the 40 workpieces. Future research should be the analysis of the rate dependence of the cutting-edge wear on the cutting time.

Author Contributions: Conceptualization, G.M.M. and N.E.G.-S.; Methodology, J.F.M.; Validation, G.A.M.-H., A.M.-S., and A.C.P.R.; Formal Analysis, M.M.M.; Investigation, J.F.M. and A.C.P.R.; Resources, M.A.M.-M. and M.M.M.; Data Curation, A.M.-S.; Writing—Original Draft Preparation, G.M.M. and N.E.G.-S.; Writing—Review & Editing, G.M.M. and N.E.G.-S.; Visualization, M.A.M.-M. and G.A.M.-H.; Supervision, G.M.M. and N.E.G.-S. All authors have read and agreed to the published version of the manuscript.

Funding: This research received no external funding.

Institutional Review Board Statement: Not applicable.

Informed Consent Statement: Not applicable.

Data Availability Statement: Data will be made available on request, contacting to the corresponding author.

Acknowledgments: The authors would like to acknowledge the invaluable contributions made by Luis Alberto Reyes Calderón from an automotive brake disk manufacturer from Puebla, Pue. Gustavo Minquiz and J. Flores acknowledges VIEP-BUAP and Tecnológico Nacional de México (Grant No. 18862.23-P).

Conflicts of Interest: The authors declare no conflict of interest.

Abbreviations

GCI	Gray Cast Iron
CGI	Compacted Graphite Iron
EDS/EDX	Energy Dispersive X-ray
As-cast	Condition of a metal casting without heat treatment
CBN	Cubic Boron Nitride
PCBN	Polycrystalline Cubic Boron Nitride
UTS	Ultimate Tensile Strength
BCBN	Binderless Cubic Bore Nitride
BUE	Build-Up Edge
BNDCC	Boron Nitride Dispersed Cemented Carbide
SEM	Scanning Electron Microscopy
AFM	Atomic Force Mircroscope
CVD	Chemical Vapor Deposition
SCGN	Ceramic inserts used in this work
CNMG	Coated carbide inserts used in this work
V_C	Cutting speed
f_n	Feed per revolution
a_P	Depth of cut

References

1. Rassini, S.A.B. de C.V. Reporte Anual 2018. 2018. Available online: https://www.bmv.com.mx/docs-pub/infoanua/infoanua_918598_2018_1.pdf (accessed on 31 July 2023).
2. Durand-Charre, M. *Microstructure of Steels and Cast Irons*; Springer: Berlin/Heidelberg, Germany, 2003.
3. Zhu, L.; Evans, R.; Zhou, Y.; Ren, F. Wear Study of Cubic Boron Nitride (cBN) Cutting Tool for Machining of Compacted Graphite Iron (CGI) with Different Metalworking Fluids. *Lubricants* **2022**, *10*, 51. [CrossRef]
4. Collini, L.; Nicoletto, G.; Konecna, R. Microstructure and mechanical properties of pearlitic gray cast iron. *Mater. Sci. Eng. A* **2008**, *488*, 529–539. [CrossRef]
5. Behnam, M.M.J.; Davami, P.; Varahram, N. Effect of cooling rate on microstructure and mechanical properties of gray cast iron. *Mater. Sci. Eng. A* **2010**, *528*, 583–588. [CrossRef]
6. Dawson, S.; Popelar, P. Thermal Analysis and Process Control for Compacted Graphite Iron and Ductile Iron. United Kingdom. 2014. Available online: https://sintercast.com/media/2280/thermal-analysis-and-process-control-for-compacted-graphite-iron-and-ductile-iron.pdf (accessed on 31 August 2023).
7. Pluphrach, G.; Teekasap, S.; Amornthatri, T.; Limboonruang, T.; Kharanan, T.; Pluphrach, K. Effects of cooling rate on impact properties and microstructure of gray cast iron ASTM A48. *Songklanakarin J. Sci. Technol.* **2022**, *44*, 929–935.
8. Hurst, S. Metal Casting—Appropriate Technology in the Small Foundry, First. Southampton Row, London: Intermediate Technology Development Group. 1996. Available online: https://www.ptonline.com/articles/how-to-get-better-mfi-results (accessed on 24 August 2023).
9. Richards, V.L.; Nicola, W. Final Technical Report: Age Strengthening of Gray Cast Iron Phase III. Produced Under Contract Number: DE-FC07-00ID13851, p. DOE/ID13851. 2003. Available online: https://www.osti.gov/servlets/purl/812004 (accessed on 31 August 2023).
10. Alexis, V.; Frédéric, R.; Jean, Q.; Eric, A. Determination of Gray Cast Iron Age Strengthening by Nondestructive Methods: Effect of Alloying Elements. *J. Mater. Eng. Perform.* **2019**, *28*, 4026–4033. [CrossRef]
11. Kountanya, R.K.; Boppana, P. Optimization of machining of automotive components with polycrystalline cubic boron nitride. *Proc. Inst. Mech. Eng. Part B J. Eng. Manuf.* **2008**, *222*, 797–805. [CrossRef]
12. Ebner, R. Influence of Aging, Sampling Site and Machining of Test Specimens Upon Tensile Strength and Hardness of Grey Cast Iron. *BCIRA* **1963**, *50*, 689–691.

13. Richards, V.L. AGE-strengthening of cast iron and its effects on machinability: Review of the literature. In *Advances in the Science and Engineering of Casting Solidification: An MPMD Symposium Honoring Doru Michael Stefanescu*; TMS Annual Meeting 2015-March; Wiley: Hoboken, NJ, USA, 2015; pp. 269–276. [CrossRef]
14. Teague, J.; Richards, V. Age strengthening of cast irons: Review of research and literature. *Int. J. Met.* **2010**, *4*, 45–57. [CrossRef]
15. Leslie, W.C. *The Physical Metallurgy of Steels*; TechBooks: MI Michigan, USA, 1981.
16. Anish, T.; Lekakh, S.N.; Richards, V.L. 08-063 The Effect of Ti and N on Iron Age Strengthening. *Trans. Am. Foundrymen's Soc.* **2008**, *116*.
17. Fall, I.; Genin, J.M.R. Mössbauer spectroscopy study of the aging and tempering of high nitrogen quenched Fe-N alloys: Kinetics of formation of Fe16N2 nitride by interstitial ordering in martensite. *Metall. Mater. Trans. A Phys. Metall. Mater. Sci.* **1996**, *27*, 2160–2177. [CrossRef]
18. Anish, T.V. Age Strengthening of Gray Cast iron: Alloying Effects and Kinetics Study. Missoury University. 2007. Available online: https://scholarsmine.mst.edu/masters_theses/4554/ (accessed on 24 August 2023).
19. Stephenson, D.A.; Agapiou, J.S. *Metal Cutting Theory and Practice, Third*; CRC Press: Boca Raton, FL, USA, 2016.
20. Smith, G.T. *Cutting Tool Technology*; Springer: London, UK, 2008.
21. de Sousa, J.A.G.; Sales, W.F.; Machado, A.R. A review on the machining of cast irons. *Int. J. Adv. Manuf. Technol.* **2018**, *94*, 4073–4092. [CrossRef]
22. Yin, G.; Shen, J.; Wu, Z.; Wu, X.; Jiang, F. Experimental Investigation on the Machinability of PCBN Chamfered Tool in Dry Turning of Gray Cast Iron. *Processes* **2022**, *10*, 1547. [CrossRef]
23. Wojciechowski, S.; Talar, R.; Zawadzki, P.; Wieczorowski, M. Evaluation of physical indicators and tool wear during grooving of spheroidal cast iron with a novel WCCo/cBN (BNDCC) inserts. *Wear* **2020**, *454–455*, 203301. [CrossRef]
24. Sayit, E.; Aslantas, K.; Çiçek, A. Tool wear mechanism in interrupted cutting conditions. *Mater. Manuf. Process.* **2009**, *24*, 476–483. [CrossRef]
25. Schultheiss, F.; Bushlya, V.; Lenrick, F.; Johansson, D.; Kristiansson, S.; Ståhl, J.E. Tool wear mechanisms of pcBN tooling during high-speed machining of gray cast iron. *Procedia CIRP* **2018**, *77*, 606–609. [CrossRef]
26. Chen, J.; Liu, W.; Deng, X.; Wu, S. Tool life and wear mechanism of WC-5TiC-0.5VC-8Co cemented carbides inserts when machining HT250 gray cast iron. *Ceram. Int.* **2016**, *42*, 10037–10044. [CrossRef]
27. Pereira, A.A.; Boehs, L.; Guesser, W.L. The influence of sulfur on the machinability of gray cast iron FC25. *J. Mater. Process. Technol.* **2006**, *179*, 165–171. [CrossRef]
28. Tooptong, S.; Park, K.H.; Lee, S.W.; Kwon, P.Y. A Preliminary Machinability Study of Flake and Compacted Graphite Irons with Multilayer Coated and Uncoated Carbide Inserts. *Procedia. Manuf.* **2016**, *5*, 644–657. [CrossRef]
29. Angseryd, J.; Coronel, E.; Elfwing, M.; Olsson, E.; Andrén, H.O. The microstructure of the affected zone of a worn PCBN cutting tool characterised with SEM and TEM. *Wear* **2009**, *267*, 1031–1040. [CrossRef]
30. Herwan, J.; Misaka, T.; Kano, S.; Sawada, H.; Furukawa, Y.; Ryabov, O. Improving Sustainability Index of Grey Cast Iron Finish Cutting Through High-Speed Dry Turning and Cutting Parameters Optimization Using Taguchi-Based Bayesian Method. *Int. J. Precis. Eng. Manuf. Green Technol.* **2022**, *10*, 729–745. [CrossRef]
31. *ISO 3685:1993*; Tool-Life Testing with Single-Point Turning Tools. International Organization for Standardization: London, UK, 1993; p. 48.
32. Tungaloyamerica. TurnLine—T515- New Grade for Cast Iron Turning. 2017. Available online: https://tungaloy.com/wpdata/wp-content/uploads/433-u_T515.pdf (accessed on 24 August 2023).
33. GmbH, C. Ceramic insert for Turning, Grooving and Milling. Germany. 2023. Available online: https://www.ceramtec.com/files/wz_ceramic_inserts_en.pdf (accessed on 24 August 2023).
34. Coromant, S. Training Handbook—Metal Cutting Technology. Swedden. 2017. Available online: https://www.sandvik.coromant.com/en-gb/search?q=C-2920-40-2&generalRefiners=%7B%7D (accessed on 31 August 2023).
35. Sobiyi, K.; Sigalas, I.; Akdogan, G.; Turan, Y. Performance of mixed ceramics and CBN tools during hard turning of martensitic stainless steel. *Int. J. Adv. Manuf. Technol.* **2015**, *77*, 861–871. [CrossRef]
36. Nicola, W.M.; Richards, V.; Edington, J. Age Strengthening of Gray Cast Iron: Nitrogen Effects and Machinability. Master's Thesis, Missouri University of Science and Technology, Rolla, MO, USA, 2002.
37. Fundamental, S. Encyclopedia of Tribology. Springer: Berlin/Heidelberg, Germany, 2013. [CrossRef]
38. Guo, X.; Chen, L.; Zhao, W.; Wan, H.; Wen, H.; Zhou, J. Machinability study on dry machining of white cast iron by polycrystalline cubic boron nitride inserts. *Mach. Sci. Technol.* **2022**, *26*, 137–159. [CrossRef]

applied sciences

MDPI

Article

Surface Quality of Al_2O_3 Ceramic and Tool Wear in Diamond Wire Sawing Combined with Oil Film-Assisted Electrochemical Discharge Machining

Zhixin Jia, Kaiyue Zhang and Jin Wang *

School of Mechanical Engineering, University of Science and Technology Beijing, Beijing 100083, China; jiazhixin@me.ustb.edu.cn (Z.J.); zhangkaiyuexks@163.com (K.Z.)
* Correspondence: wangjin84@ustb.edu.cn

Abstract: Diamond wire sawing is one of the most widely used methods of cutting Al_2O_3 ceramic because it has good machining accuracy and causes less surface damage. However, its material removal rate (MRR) needs to be improved with the increasing demand for Al_2O_3 ceramic parts. In this paper, spark discharges are generated around the diamond wire based on the electrochemical discharge machining (ECDM) process. An oil film-assisted ECDM process is applied to solve the difficulty of generating spark discharges when the thickness of the workpiece exceeds 5.0 mm due to the difficulty of forming a hydrogen gas film. Experimental results show that the combination of oil film-assisted ECDM and diamond wire sawing improved the MRR of Al_2O_3 ceramic. Oil film-assisted ECDM may improve the surface quality of machined parts and reduce the wear on diamond wire. Therefore, this research focuses on the surface quality of Al_2O_3 ceramic and tool wear in diamond wire sawing combined with oil film-assisted ECDM. Surface roughness and topography, recast layer, and elements of the machined surface are analyzed. The tool wear is studied using SEM images of diamond wire. The results provide a valuable basis for application of diamond wire sawing combined with oil film-assisted ECDM.

Keywords: surface quality; tool wear; oil film-assisted ECDM; diamond wire sawing; Al_2O_3 ceramic

check for updates

Citation: Jia, Z.; Zhang, K.; Wang, J. Surface Quality of Al_2O_3 Ceramic and Tool Wear in Diamond Wire Sawing Combined with Oil Film-Assisted Electrochemical Discharge Machining. *Appl. Sci.* **2023**, *13*, 9030. https://doi.org/10.3390/app13159030

Academic Editors: Rui Manuel Leal, Gustavo H. S. F. L. Carvalho and Ivan Galvão

Received: 30 June 2023
Revised: 31 July 2023
Accepted: 2 August 2023
Published: 7 August 2023

1. Introduction

Al_2O_3 ceramic is an important engineering material because of its high hardness, electrical resistivity, strength, wear resistance and low density. However, its hard and brittle nature and electrically insulating characteristic make machining difficult and costly. Diamond wire sawing is one of the most effective ways of cutting Al_2O_3 ceramic because it has good machining accuracy and causes less surface damage [1]. However, the MRR needs to be improved due to the increasing demand for Al_2O_3 ceramic parts. In this paper, an oil film-assisted ECDM process is integrated with diamond wire sawing. The high temperature generated from spark discharge lowers the strength of Al_2O_3 ceramic and makes it easier to be removed by the moving diamond wire.

The integration of oil film-assisted ECDM may bring about changes in surface quality and affect the performance of machined parts. At present, the main methods of electrical discharge machining of ceramic materials are EDM and ECDM. Some studies have added conductive materials to Al_2O_3 ceramics so that electric discharge machining (EDM) can be used. Lin [2] et al. prepared MgF_2-doped Al_2O_3 ceramics by the stirring and mixing method and non-pressure sintering process, effectively improving the dielectric constant of Al_2O_3 ceramics. Takayuki et al. in Japan studied the phenomenon of wire-cutting of insulating ceramics by covering the surface of insulating ceramics with a layer of conductive material using the PVD method, and then using molybdenum, copper or copper zinc electrode wires for wire cutting. Takayuki also produced complex axisymmetric shapes by rotating the workpiece [3]. Wu et al. added conductive powder into the electrolyte and found that the surface of hard and brittle non-conductive materials can be processed efficiently and better surface quality can be obtained by mixing powder electrolytic and EDM technology [4]. Patel et al. [5] studied the surface quality of Al_2O_3 ceramic composite reinforced by TiC particles and SiC whiskers during EDM, and the results show that the discharge energy can cause thermal spalling, which greatly increases the MRR. The above is a study on the surface quality of conductive Al_2O_3 ceramic composite materials by EDM. When the workpiece material does not conduct electricity, do not use EDM to generate spark discharge. Electrochemical discharge machining (ECDM) is a solution to generate spark discharge for machining insulation materials [6], although its MRR is extremely low when machining Al_2O_3 ceramics [7]. Lijo et al. [8] processed semi-conductive engineering materials with mixed electrolyte of sodium hydroxide and potassium hydroxide during ECDM and found that the mixed electrolyte improved the processing speed in terms of material removal rate. Liu et al. [9] found the optimal combination of process parameters, which achieved better film quality and more uniform discharge energy distribution, thus reducing radial overcutting and roundness errors during ECDM. Wang et al. [10] designed and developed a tracking electrode device to make it move synchronously with the tool tube electrode, thereby stabilizing the electrode between the anode and cathode to improve the machining stability. This improved the machining effect and the surface quality of the micro-hole of insulated ceramic ECDM. Cao et al. [11] proposed that laser-assisted ECDM could significantly improve the quality and precision of electrochemical discharge machining microchannels, aiming at the existing problems of heat-affected zone and over-cutting phenomenon in electrochemical discharge machining. Based on the wetting method, Shi et al. [12] designed a fuzzy control system based on force signal feedback control of feed speed. Applying it to the machining experiment platform of force signal feedback control of feed effectively improved the overcutting phenomenon on both sides of the micro-groove during ECDM. When ECDM micro-drilling electrical insulating glass materials, micro-cracks caused by electrochemical discharge were found on the machined surface [13–15] and the heat-affected zone was observed near the machined surface [16]. Singh et al. [17] used the TOPSIS method to optimize the process parameters of ECDM, improve the overcutting phenomenon of processed samples, and improve the surface quality of processed samples, but there were still micro-cracks on the surface of the samples. In conclusion, high spark discharge temperature usually has a negative impact on the machined surface. Therefore, it is necessary to analyze the surface quality of Al_2O_3 ceramic in diamond wire sawing combined with oil film-assisted ECDM.

The integration of oil film-assisted ECDM will also bring changes to the wear of the diamond wire. In recent years, electrochemical machining technology has become a potential competitor in the processing of non-conductive hard and brittle materials, but due to the presence of abrasive particles, the conventional processing of these materials will lead to high tool wear [18]. Khan [19] analyzed tool wear during EDM of conductive Al_2O_3 composites. In the process of EDM, the wear rate of electrode increases with the increase of applied current and voltage. Sumit et al. [20] found that lower electrolyte concentrations reduced overcutting by 22%, as well as tool wear and hole taper by 39% and 18%, respectively. Kumar et al. [21] found that nickel plating on cutting tools can reduce the average overcut and heat-affected zone width, and also improve the overall roundness of machining micro-holes. Bian et al. [22] analyzed the influence mechanism of machining parameters on tool wear in the non-machining state and obtained the voltage range suitable for machining. Based on the current signal, it was found that the local material of the cathode melted or vaporized at high temperature, and this loss behavior may shorten the film forming time, but has little effect on the average spark discharge current. Xie [23] measured the external profile of the tool electrode using the contact method and found that negative polarity machining, high polarity voltage, small peak current, large gain, high frequency pulse and long pulse width can reduce wear on the length direction of the tool electrode. In Japan, a thin metal mesh was pressed on the wafer workpiece as an auxiliary electrode for processing non-conductive ceramic materials. The vaporization, spraying or sputtering of metal materials was used to make the surface of ceramic parts conductive, so that the processing could be sustained. However, the processing depth of this method was relatively shallow [24–26]. Experimental investigations were carried out on the performance of electrochemical discharge in machining glass–epoxy and Kevlar–epoxy composites [27]. The results show that the tool wear increases with the increase of voltage. These research studies support that the spark discharges have an influence on tool wear. Therefore, the research on tool wear in diamond wire sawing combined with oil film-assisted ECDM is necessary.

In this research, we conducted a series of experiments to analyze the surface quality of Al_2O_3 ceramic and tool wear in diamond wire sawing combined with oil film-assisted ECDM. Surface roughness and topography, recast layer and elements of the machined Al_2O_3 ceramic were investigated. The surface topography of used diamond wire was observed to analyze the wear of diamond wire.

2. Materials and Methods

2.1. Oil Film-Assisted ECDM

ECDM wire cutting is a kind of cutting method for hard and brittle insulating materials which has been increasingly studied. The processing principle is shown in Figure 1. Generating a complete hydrogen film on the electrode wire surface is one of the necessary conditions for ECDM discharge [28]. However, due to the combined effect of the buoyancy of the electrolyte and the motion of the electrode wire, the hydrogen generated by the electrochemical reaction easily leaves the surface of the electrode wire, which brings difficulties to the formation of the hydrogen film. With the increase of cutting thickness, the length of electrode wire required to generate discharge increases correspondingly, and the length of electrode wire required to be covered by hydrogen film also increases. Hydrogen film formation is, therefore, more difficult, and even if it can be formed, it is extremely vulnerable to damage, resulting in difficult or even impossible discharge formation. In addition, the need to avoid deterioration of the hydrogen film makes it difficult to use effective means (such as forced flushing) to improve the electrolyte circulation in the processing area. Therefore, the discharge stability of ECDM machining is poor, the processing efficiency is low, and only two-dimensional cutting processing of hard and brittle insulating materials with small thickness (about 1.0 mm) can be achieved at present. Therefore, hydrogen film formation is difficult and it is easily destroyed, which is an urgent problem to be solved in ECDM wire cutting technology.

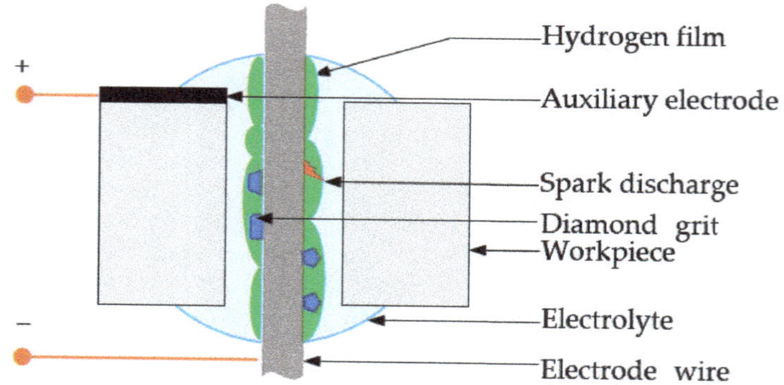

Figure 1. Schematic diagram of electrical discharge wire cutting of insulating materials.

It is difficult to solve the problem that hydrogen film formation is difficult and it is easily destroyed in ECDM. In this paper, an oil coating device is used to coat a uniform oil film on the electrode wire in advance to replace the hydrogen film. Thus, the electrical insulation conditions between the electrode wire and the electrolyte can be met, and then discharge can be generated to remove the workpiece material. This is the basic principle of the new technology of oil film-assisted ECDM machining proposed in this paper. As shown in Figure 2a, the electrode wire is first coated with oil by the oiling device in the air, and a complete oil film is formed on the surface. Then, the electrode wire enters the electrolyte. Due to the existence of the oil film on the surface, the electrode wire and the electrolyte are in a state of electrical insulation. No electrochemical reaction occurs, and the current in the circuit is zero. At this time, the main function of the electrolyte is to conduct the potential of the anode (auxiliary electrode) to the vicinity of the oil film, thus forming a potential difference on both sides of the oil film, the size of which is equal to the output voltage of the pulse power supply. When the voltage exceeds the critical breakdown voltage of the oil film, a spark discharge is formed between the electrode wire and the electrolyte. The potential difference between the electrode wire and the electrolyte is suddenly reduced, the discharge current is generated in the circuit, and the workpiece material is removed under the high temperature condition of discharge. As shown in Figure 2b, after discharge the oil film at the discharge position will be damaged by high temperature and have a slight defect. When the electrode wire contacts with the electrolyte, an electrochemical reaction will occur to produce hydrogen. Therefore, after discharge, the insulating film on the surface of the electrode wire is an oil–hydrogen two-phase dielectric film with oil film as the main layer and hydrogen film as the auxiliary layer. The two-phase dielectric film is in a state of "breakdown discharge—damage—electrochemical reaction repair—breakdown discharge...". The distribution of the cyclic process is constantly dynamically evolving, and the processing state (which can include breakdown delay, oil film breakdown and discharge, damaged oil film repair, gas film breakdown and discharge, as well as arc and electrode wire contact with the workpiece) is also constantly changing. Therefore, oil film-assisted ECDM is a new technology of hard and brittle insulation material processing, which can remove the workpiece material by continuous discharge under the influence of oil film forming, dynamic evolution of insulation film distribution and various processing states.

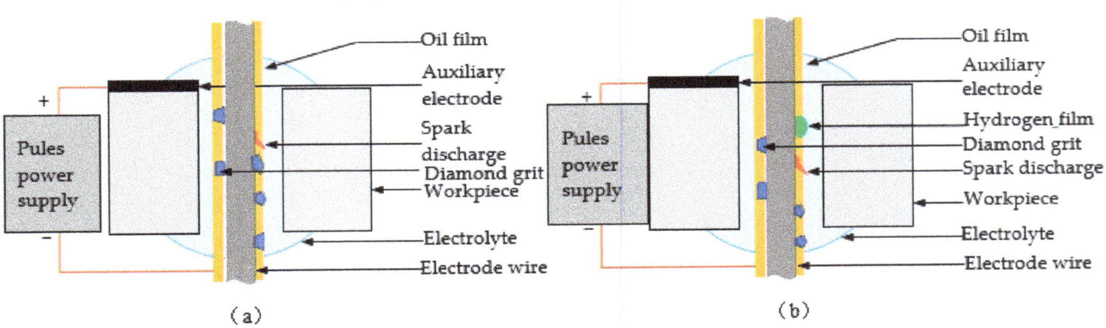

Figure 2. Principle of spark discharge generation by oil film-assisted ECDM during diamond wire sawing: (**a**) The oil film on the electrode wire surface is complete, (**b**) The oil film on the electrode wire surface is defective.

2.2. Feasibility Analysis

The oil film-assisted ECDM technology proposed in this paper has both gas film and liquid film in the process. In order to verify whether the spark discharge between the electrode wire and the electrolyte can breakdown the oil film, the experimental device shown in Figure 3 is designed by cleverly taking advantage of the fact that the density of oil is less than that of water.

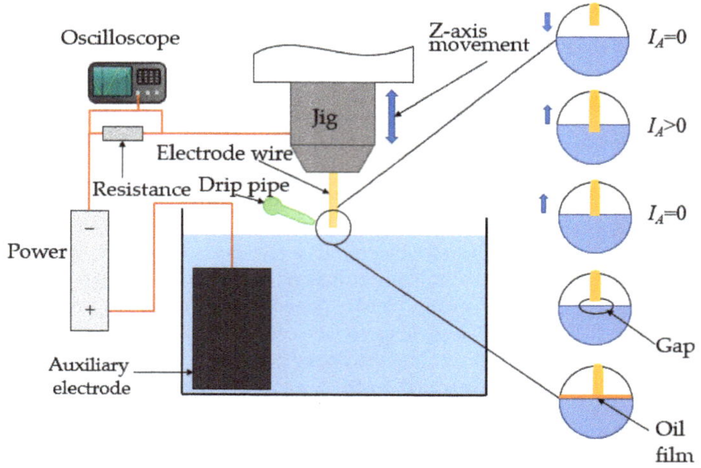

Figure 3. Schematic diagram of breakdown voltage measurement principle of complete oil film covered by electrode wire.

The electrolyte is placed in a transparent glass beaker, the auxiliary electrode is placed at the bottom of the beaker, and the electrode wire is installed on the fixture of the vertical feeding mechanism. A small voltage is applied between the electrode wire and the auxiliary electrode, while the feed mechanism drives the electrode wire straight down. In this process, the current in the circuit is measured in real time. When the electrode wire does not drop to the electrolyte level in the beaker, the current in the circuit is zero, and when the electrode wire contacts the electrolyte level, the current is generated in the circuit. Thereafter, the feed mechanism drives the electrode wire gradually up at a slow speed until the current in the circuit is just zero, at which time the electrode wire is just free from the electrolyte. Based on this position, the electrode wire continues to move upward to form a gap between the electrode wire and the electrolyte surface, and the distance of the electrode wire moving upward is equal to the size of the gap. Slowly drop the oil into the electrolyte level to ensure that the oil fills the gap between the electrode wire and the electrolyte. At this time, the thickness of the oil between the electrode wire and the electrolyte is equivalent to the thickness of the oil film covering the surface of the electrode wire, and the oil film thickness is equal to the gap distance between the electrode wire and the electrolyte. The voltage between the electrode wire and the auxiliary electrode is gradually increased, the voltage and current signals are collected, and the volt–ampere characteristic curve is established to determine the breakdown voltage threshold.

Figure 4a shows the experimental device, which is carried out on a high-speed EDM perforator. The electrode wire is connected to the negative electrode of the power supply, and the auxiliary electrode is connected to the positive electrode of the power supply. Figure 4b shows the spark discharge between the end face of the electrode wire and the electrolyte after the voltage is applied. The experiment can prove that the spark discharge between the electrode wire and the electrolyte can indeed breakdown the oil film.

To sum up, the technology of diamond wire sawing combined with oil film-assisted ECDM proposed in this paper is practical and feasible, which can solve the problem of poor stability or even impossibility of traditional EDM, improve processing efficiency and processing quality, and thus solve the processing problems of hard and brittle insulating materials. At the same time, this method significantly reduces the energy consumption for maintaining the electrochemical reaction to generate hydrogen. It has obvious advantages in energy conservation.

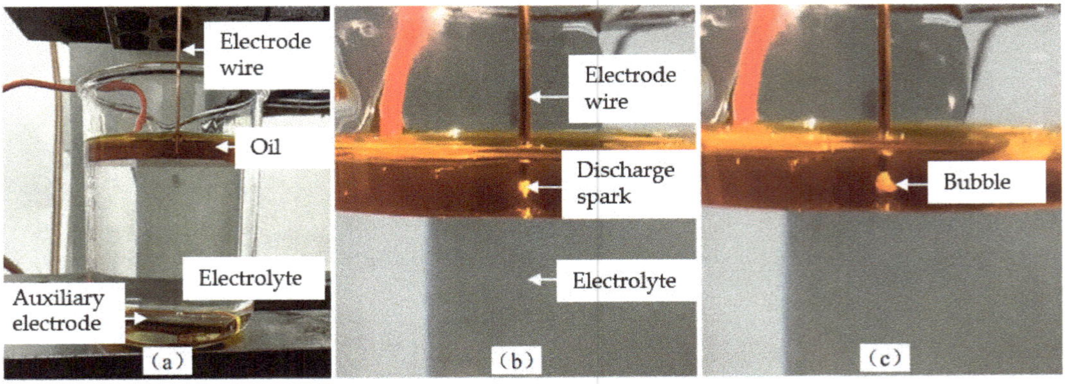

Figure 4. Test of breakdown oil film discharge between electrode wire and electrolyte: (**a**) Experimental apparatus; (**b**) Spark discharge; (**c**) Bubbles formed by electrical discharge.

2.3. MRR Improvement

The equipment used in this study is obtained through the transformation of DK7720 fast wire EDM wire-cutting machine, mainly replacing the molybdenum wire used in the EDM wire-cutting machine with a diamond wire saw, and adding the discharge circuit and electrolyte supply device to the machine. The movement of the diamond wire is driven by a rotating roller, and the workpiece is fixed on a slider mounted on the guide rail. The slider is pulled by the counterweight and moves forward as the working material is removed. In this feed mode, the workpiece is only moved forward when the working material is removed. The feed rate of the workpiece is not a process parameter, but a result of the material removal of the workpiece [29].

Experiments were conducted to analyze the MRR of Al_2O_3 ceramic by diamond wire sawing with and without the assistance of oil film-assisted ECDM. The characteristics of the Al_2O_3 ceramic are shown in Table 1. The guide rail oil with properties shown in Table 2 was used to online form an oil film on the diamond wire. The experimental conditions are shown in Table 3. An NaCl solution with a mass fraction of 20% was chosen as the electrolyte because of its non-corrosive and good electrical conductivity properties, and the velocity of spraying NaCl solution was 0.583 m/s. Figure 5 shows the experimental picture of cutting Al_2O_3 ceramic by diamond wire sawing combined with oil film-assisted ECDM. It can be seen that spark discharges were generated around the diamond wire (Figure 5a), and a cutting slit was produced (Figure 5b). Figure 6 shows that the diamond wire sawing combined with oil film-assisted ECDM increased MRR by 0.46 mm^3/min or 36%. MRR is approximated by the formula: MRR = slit length × slit width × workpiece thickness/processing time, where slit length is measured by a microscope, and slit width is approximately the diameter of a wire saw.

Table 1. Characteristics of the Al_2O_3 ceramic.

Density	Hardness	Melting Point	Electrical Resistivity	Thermal Conductivity
3.9 g/cm^3	9 (Mohs)	2054 °C	10^{16} Ω cm	35 W/m K

Table 2. Properties of the guide rail oil used to form oil film on the diamond wire.

Density	Kinematic Viscosity	Flash Point	Pour Point
0.87–0.89 g/cm^3	69.24 mm^2/s (40 °C)	233 °C	−21 °C

Table 3. Experimental conditions of analyzing MRR of Al_2O_3 ceramic by diamond wire sawing with and without the assistance of oil film-assisted ECDM.

Conditions	Diamond Wire Sawing	Spark Discharge-Assisted Diamond Wire Sawing
Diamond wire diameter	0.2 ± 0.01 mm	0.2 ± 0.01 mm
Workpiece thickness	10.0 mm	10.0 mm
Working liquid	Water	NaCl solution (20%)
Auxiliary electrode	-	Stainless steel plate
DC voltage	-	52 V
Wire speed	1400 mm/s	1400 mm/s
Counterweight mass	400 g	400 g
Machining time	5 min	5 min

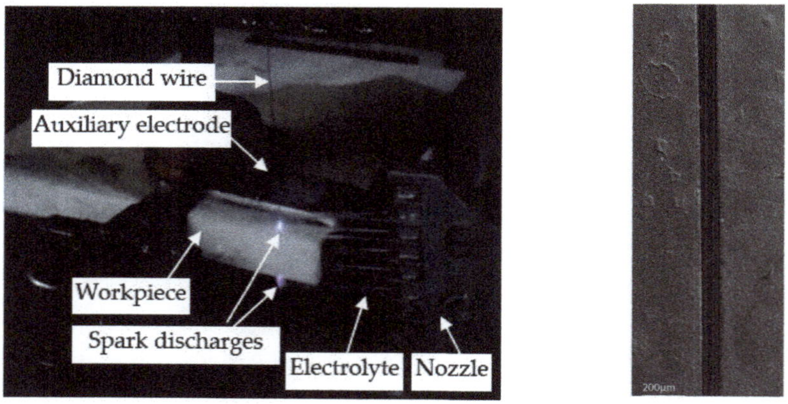

(a) (b)

Figure 5. Experimental pictures of cutting Al_2O_3 ceramic by diamond wire sawing combined with oil film-assisted ECDM: (**a**) Cutting process, (**b**) Produced slit.

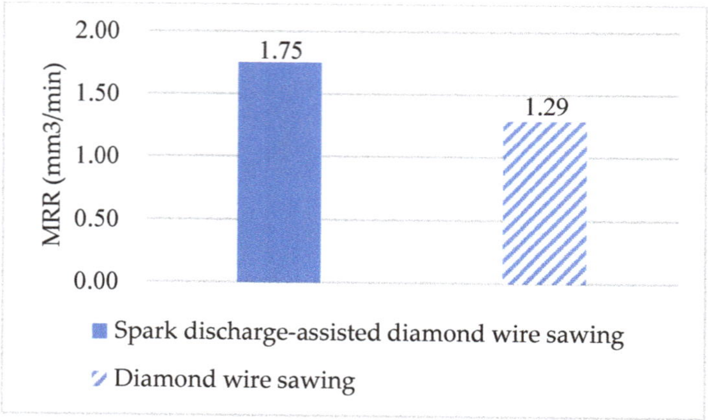

Figure 6. MRR of Al_2O_3 ceramic by diamond wire sawing with and without the assistance of oil film-assisted ECDM.

2.4. Experiments on Surface Quality and Tool Wear

Experiments were conducted to analyze the surface quality of Al_2O_3 ceramic by diamond wire sawing combined with oil film-assisted ECDM. Samples with dimensions were cut off from an Al_2O_3 ceramic plate using diamond wire sawing combined with oil film-assisted ECDM under different DC voltages, and other conditions are the same as Table 3. The length, width and height of the sample sizes were 10, 15 and 5 mm, respectively. As a comparative object, a sample with the same dimensions was obtained by diamond wire sawing.

Surface roughness was measured by a surface roughness tester. Surface topography was observed through scanning electron microscope (SEM). Because the Al_2O_3 ceramic is non-conducive, a layer of carbon was applied on the samples by a high vacuum evaporator before SEM. Each sample has one pre-polished surface, so the recast layer can be observed through SEM. Energy dispersive X-ray spectroscopy (EDS) was applied to analyze the elements' variation on the machined surface.

Surface topography of diamond wires after cutting Al_2O_3 ceramic with and without the assistance of oil film-assisted ECDM were observed using SEM. The wear of diamond wire was analyzed according to these SEM images.

3. Results and Discussion

3.1. Surface Roughness and Topography

This part of the study explores the influence of the addition of the oil film-assisted ECDM process on the cutting surface roughness, which needs to reflect the changing trend of the surface roughness value under different voltages, and then compare the roughness value with or without the assistance of discharge process. There are many evaluation parameters of roughness, and selecting one evaluation parameter is convenient for comparison. It is generally believed that Ra is most commonly used to evaluate roughness and can fully reflect the surface micro-geometric characteristics. Therefore, Ra is selected as the parameter for evaluating roughness in this paper. The surface roughness was measured by a portable metal surface roughness tester, TR100. Surface roughness of the machined samples was measured in two directions: one was parallel to the diamond wire, the other was perpendicular to the diamond wire. After three measurements, when no voltage was applied, the surface roughness of the machined sample by diamond wire sawing in the parallel direction was, respectively, 0.441 μm, 0.452 μm, 0.463 μm, and the average value was 0.452 μm; the perpendicular direction was, respectively, 0.597 μm, 0.692 μm, and 0.616 μm, and the average value was 0.627 μm. When diamond wire sawing combined with oil film-assisted ECDM, surface roughness values in the parallel direction and perpendicular direction under different voltage applications are shown in Table 4. The relationship between surface roughness and applied DC voltage is shown in Figure 7. It can be seen that the surface roughness Ra achieves the minimum value of 0.485 μm in the parallel direction and 0.732 μm in the perpendicular direction at 46 V. Compared with the direct cutting of diamond wire saw, the roughness increases by 7.30% in the parallel direction and 16.7% in the perpendicular direction after 46 V is applied. It also shows that the surface roughness increases with the voltage. When the applied voltage is 58 V, the roughness in the parallel direction and the perpendicular direction increases by 83.5% and 36.9%, respectively, compared with that when the applied voltage is 46 V. The results support that the integration of spark discharge with diamond wire sawing increases surface roughness. The reason is that the vibration of the diamond wire causes cutting lines on the machined surface (Figure 8).

SEM images of the machined surfaces by diamond wire sawing with and without the assistance of oil film-assisted ECDM are shown in Figure 9. During diamond wire sawing, the material was removed by the ductile regime method and spalling of work material left micro craters on the machined surface [1] (Figure 9a). Figure 9b shows the machined surface by diamond wire sawing combined with oil film-assisted ECDM under a voltage of 49 V. Compared to Figure 9a, the ductile regime area reduced, and the number of micro craters increased. When the voltage increased to 55 V, the craters became predominant on the machined surface (Figure 9c). With a DC voltage of 58 V, the craters became so dense that it is difficult to identify whether the ductile regime existed on the machined surface (Figure 9d). The SEM images indicate that the increase of DC voltage improves the spalling of Al_2O_3 ceramic. The reason is that the conflicting compressive and tensile stresses were generated by the spark discharges and the cooling electrolyte, which also existed in EDM of conductive Al_2O_3 ceramic composites [30]. The severe stress differential reduced the strength of the Al_2O_3 ceramic, which facilitated the spalling of Al_2O_3 ceramic by the moving diamond wire. The spalling of Al_2O_3 ceramic produces micro craters on machined surfaces, resulting in increased surface roughness. In addition, the increase of MRR was also due to the spalling of the Al_2O_3 ceramic. The reason is that the brittle materials are removed by plastic deformation and spalling during the grinding process, and the MRR associated with spalling is an order of magnitude greater than that for plastic deformation [31]. No fusing structure like irregular and spherical droplets was present on the machined surface, which means that the material removed by diamond wire sawing combined with oil film-assisted ECDM was due to mechanical failure instead of melting.

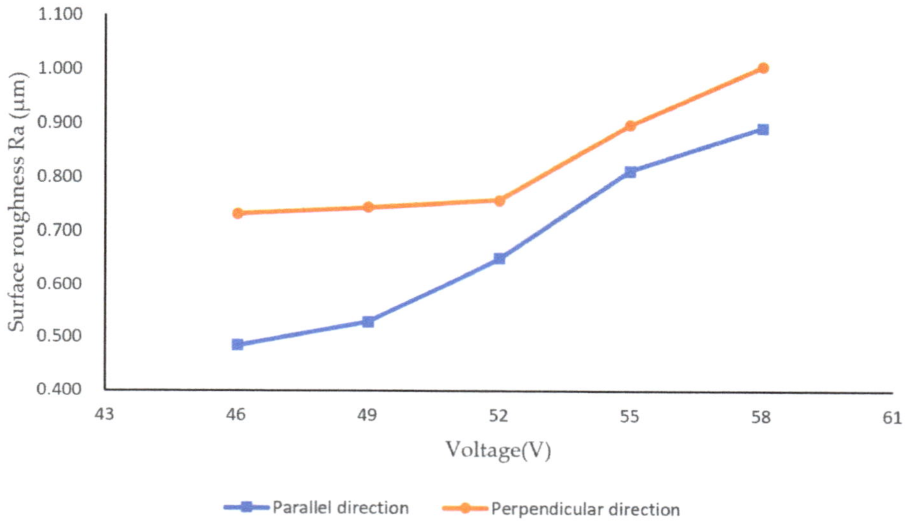

Figure 7. Effect of voltage on surface roughness.

Table 4. Surface roughness of oil film-assisted ECDM.

Measurement Direction	Voltage	First Measurement	Second Measurement	Third Measurement	Mean Value	Standard Error
Parallel	46	0.454	0.552	0.448	0.485	0.033711
	49	0.502	0.583	0.501	0.529	0.027168
	52	0.594	0.674	0.679	0.649	0.027538
	55	0.837	0.750	0.850	0.812	0.031392
	58	0.865	0.838	0.966	0.890	0.038954
Perpendicular	46	0.756	0.681	0.760	0.732	0.025693
	49	0.714	0.802	0.714	0.743	0.029333
	52	0.751	0.711	0.810	0.757	0.028754
	55	0.869	0.869	0.950	0.896	0.027000
	58	0.939	1.018	1.050	1.002	0.032987

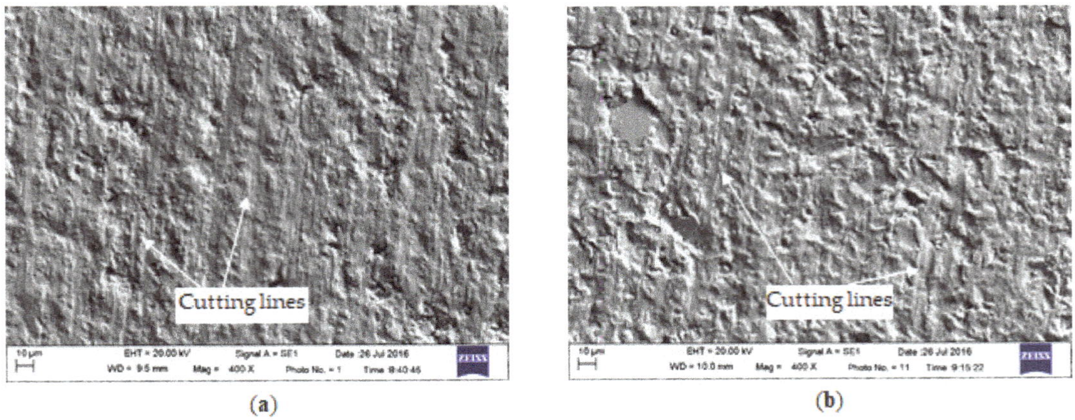

Figure 8. SEM of machined surface by (**a**) diamond wire sawing, and (**b**) diamond wire sawing combined with oil film-assisted ECDM (46 V).

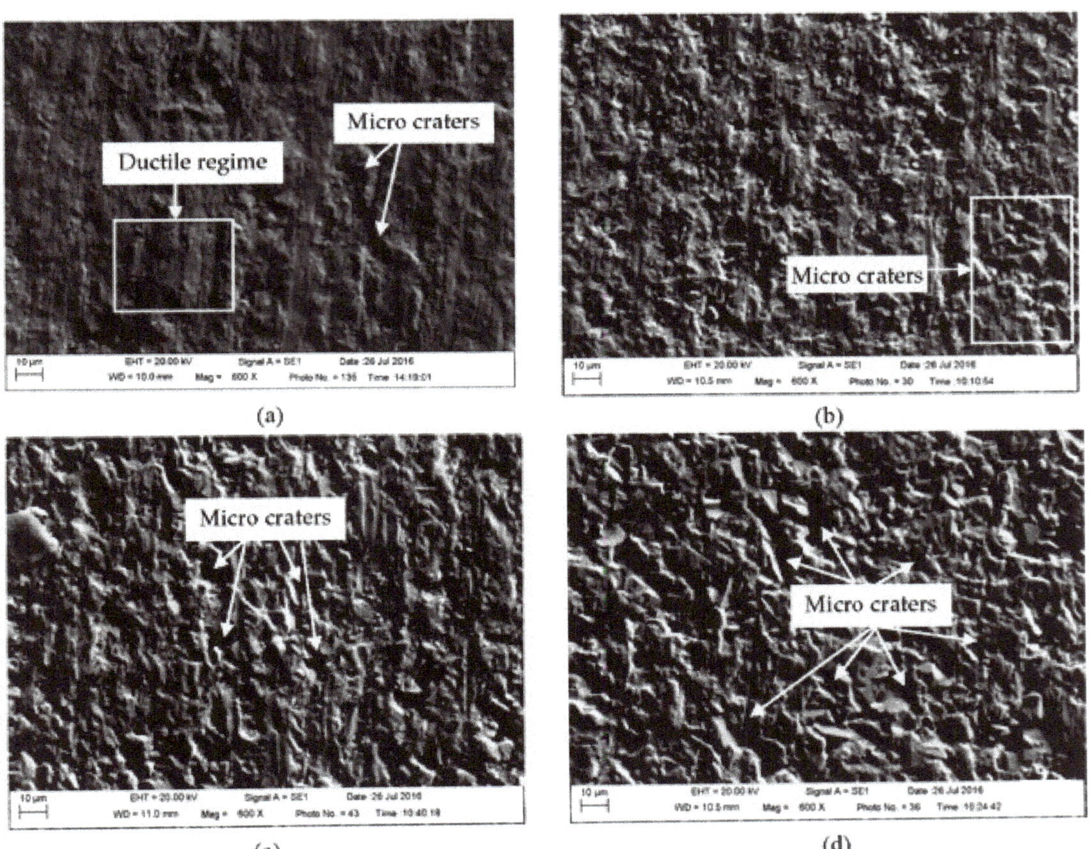

Figure 9. SEM of machined surfaces by (**a**) diamond wire sawing, and diamond wire sawing combined with oil film-assisted ECDM under DC voltages of (**b**) 49 V, (**c**) 55 V, and (**d**) 58 V.

It is known that material removal of advanced ceramics is due to the combined action of melting, evaporation and spalling. Furthermore, in ceramic materials, thermal spalling is due to sudden changes in temperature that lead to rapid shrinkage and expansion of the material, resulting in thermal stress. This causes tensile and compressive stresses sufficient to cause the material to suddenly break, resulting in the material being removed in the form of flakes [32]. There is no micro crack on the machined surfaces when DC voltage is applied and diamond wire sawing is assisted by oil film, which is different from the machined surface by EDM. An SEM image at high magnitude was obtained to make sure that no micro crack is presented on the machined surface (Figure 10). The reason could be that the spark discharge energy is insufficient to melt the Al_2O_3 ceramic. The stresses caused by the quick variation of temperature are not large enough to generate micro cracks on the machined surface.

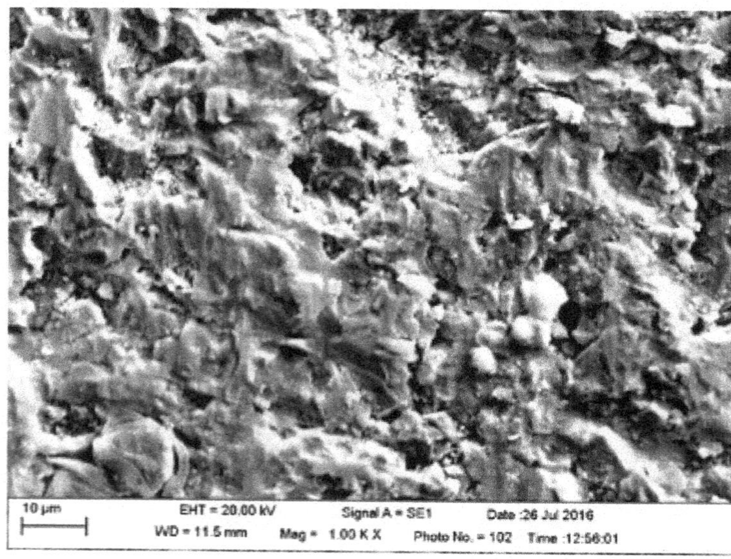

Figure 10. SEM of machined surface by diamond wire sawing combined with oil film-assisted ECDM (58 V).

3.2. Recast Layer

The recast layer was analyzed by SEM images, and the results are shown in Figure 11. SEM of a machined surface by diamond wire sawing (Figure 11a) shows that no recast layer is presented on the machined surface, but the spalling of the Al_2O_3 ceramic was found on the profile of the cutting slit. Figure 11b shows the SEM of machined surface by diamond wire sawing combined with oil film-assisted ECDM. There was no evidence supporting the existence of recast layer on the machined surface, and the spalling of Al_2O_3 ceramic was found on the profile of the cutting slit. As the voltage increased, the spalling of Al_2O_3 ceramic became severe, but there was still no recast layer on the machined surface (Figure 11c,d). The absence of recast layer is due to the low energy of spark discharges by oil film-assisted ECDM.

3.3. Elements

During EDM, the tool material can be transferred to the workpiece surface in melting and gaseous states simultaneously [33,34]. In spark discharge-assisted diamond wire sawing, the material of diamond wire can be left on the machined surface if the transferring process exists. The diamond wire is covered by Ni, which is used for fixing diamond grits, therefore, Ni is most likely to be transferred to the machined surface. Figures 12 and 13 show the EDS results of machined surfaces by diamond wire sawing with and without the assistance of oil film-assisted ECDM, respectively. It can be seen that Na and Cl were presented on the machined surface when spark discharge was integrated with diamond wire sawing. These elements came from the NaCl solution which was not totally cleaned after machining due to the micro craters on the machined surface. Nickel was not found on the machined surface by spark discharge-assisted diamond wire sawing. The reason is that the spark discharges can only melt a small amount of diamond wire materials, and most of the melting materials condense into solid particles before reaching the surface of the workpiece.

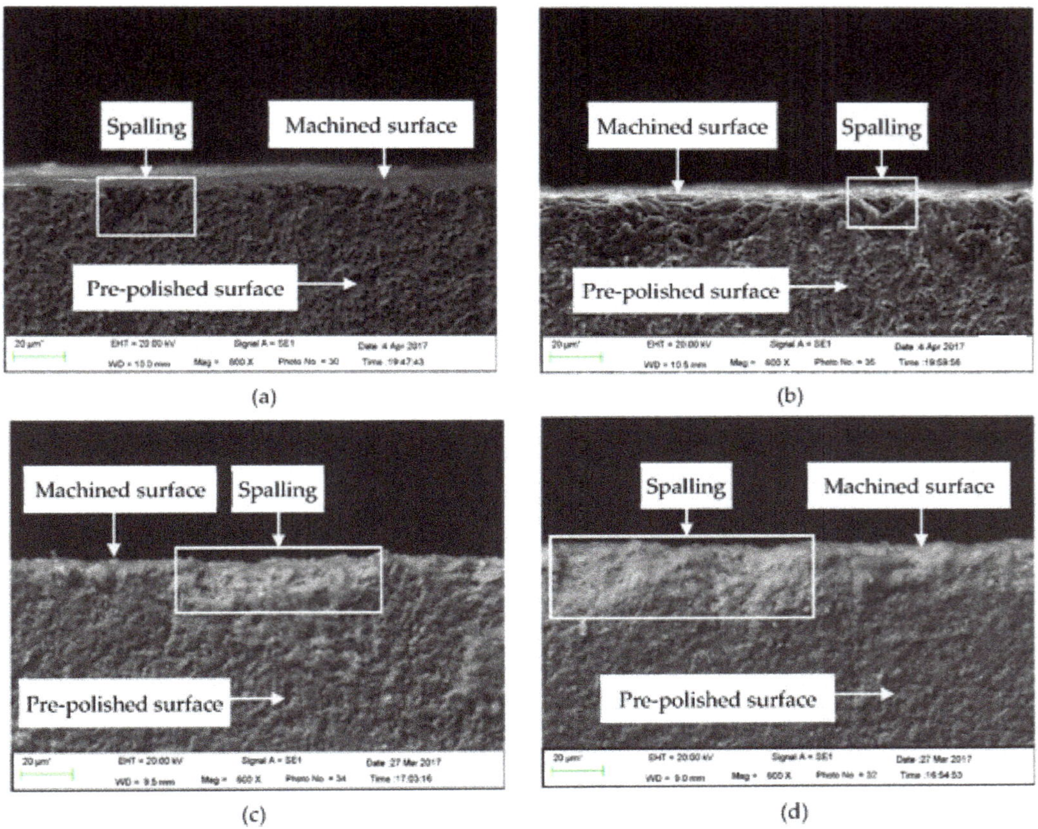

Figure 11. SEM of machined surfaces by (**a**) diamond wire sawing, and diamond wire sawing combined with oil film-assisted ECDM under DC voltage of (**b**) 46 V, (**c**) 52 V, and (**d**) 58 V.

Coating the electrode wire surface with oil film can promote discharge, but the high temperature generated by discharge may also cause the pyrolysis or even decomposition of kerosene molecules to produce carbon attached to the surface of the workpiece, affecting the removal rate of the workpiece material. Through the analysis of the element content of the processed surface, it can be found that there is no accumulation of carbon elements on the surface of the workpiece, that is, the temperature generated by the discharge is not enough to decompose the carbon in the kerosene, so it will not affect the removal rate of the workpiece material.

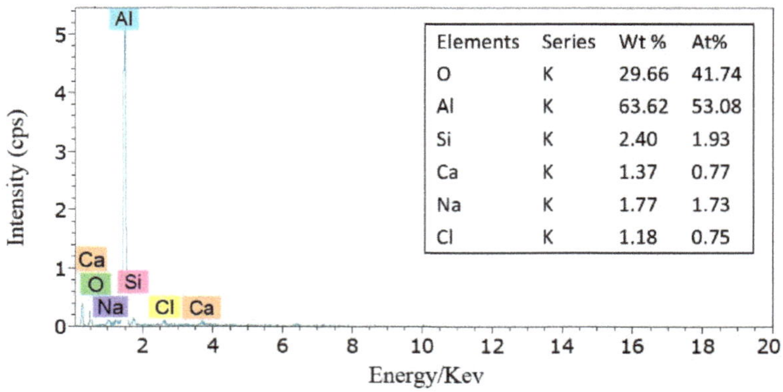

Figure 12. EDS showing the relative intensities of various elements on the machined surface by spark discharge-assisted diamond wire sawing (58 V).

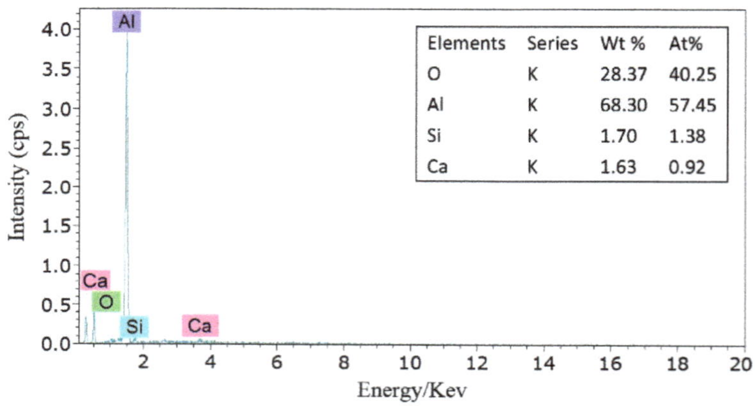

Figure 13. EDS showing the relative intensities of various elements on the machined surface by diamond wire sawing.

3.4. Tool Wear

Surface topography of the diamond wires after diamond wire sawing with and without the assistance of oil film-assisted ECDM are shown in Figure 14. It can be seen that many micro craters were produced on the matrix of diamond wire (Figure 14a), but they were not present on the diamond wire used in diamond wire sawing (Figure 14b). The micro craters were caused by material removal due to the high temperature generated from spark discharges. Figure 14a shows that one entire diamond grit was pulled out of the diamond wire matrix and a cavity was produced. The reason is that partial nickel material used to fix the diamond grit was removed by the spark discharges, resulting in the decrease of bonding strength. Actually, the diamond pull-out also exists on diamond wire after diamond wire sawing [35], but the integration of spark discharges may increase the chance of diamond grit pull-out.

(a) (b)

Figure 14. SEM of diamond wires after (**a**) diamond wire sawing combined with oil film-assisted ECDM (58 V) and (**b**) diamond wire sawing.

4. Conclusions

The integration of spark discharge with diamond wire sawing is proved to be effective in improving MRR of Al_2O_3 ceramic. Surface quality of Al_2O_3 ceramic and wear of diamond wire in diamond wire sawing combined with oil film-assisted ECDM were analyzed and the conclusions were drawn as follows:

(1) The integration of spark discharge with diamond wire sawing facilitates the spalling of Al_2O_3 ceramic, which results in the increase of surface roughness. The roughness in parallel and perpendicular directions increases by 7.30% and 16.7%, respectively when 46 V is applied. The surface roughness increases with the applied DC voltage and when the applied voltage is 58 V, the roughness increases by 116% and 38.2%, respectively.

(2) There are no recast layer and micro cracks on the machined surface by diamond wire sawing combined with oil film-assisted ECDM. The material removal is due to mechanical failure instead of melting.

(3) The Na and Cl elements are left on the machined surface, but the material of the diamond wire does not transfer to the machined surface. Through analysis of the element content on the surface of the workpiece, it can be concluded that the oil film will not decompose to produce a carbon layer attached to the surface of the workpiece, which will have a negative impact on the material removal rate of the workpiece.

(4) Micro craters are formed on the matrix of diamond wire, and the removal of Ni material increases the chance of diamond grits pull-out.

Author Contributions: Conceptualization, Z.J.; writing, K.Z.; methodology, Z.J.; writing—original draft preparation, K.Z.; data curation, J.W.; supervision, Z.J.; project administration, J.W.; funding acquisition, J.W.; investigation, J.W. All authors have read and agreed to the published version of the manuscript.

Funding: This research was funded by the National Natural Science Foundation of China (52275400); Fundamental Research Funds for the Central Universities (FRF-TP-19-003A3).

Institutional Review Board Statement: Not applicable.

Informed Consent Statement: Not applicable.

Conflicts of Interest: The authors declare no conflict of interest.

References

1. Teomete, E. Effect of process parameters on surface quality for wire saw cutting of alumina ceramic. *Gazi Univ. J. Sci.* **2011**, *24*, 291–297.
2. Li, C.Y.; Yuan, L.; Li, W. Effect of MgO/MgF_2 doping on sintering, microstructure and microwave dielectric properties of Al_2O_3 ceramics. *Bull. Chin. Ceram. Soc.* **2019**, *12*, 45–48.
3. Xu, S.L. Development Status of EDM Technology in Japan. *Electr. Machin. Tool.* **2001**, *3*, 13–16.
4. Wu, J.J.; Gu, A.; Lv, C.W.; Gan, J.B. Research on Mixed Powder Electrochemical Discharge Composite Machining Technology. *Electromachin. Mould* **2012**, *300*, 64–66.
5. Patel, K.M.; Pandey, P.M.; Rao, P.V. Surface integrity and material removal mechanisms associated with the EDM of Al_2O_3, ceramic composite. *Int. J. Refract. Met. Hard Mater.* **2009**, *27*, 892–899. [CrossRef]
6. Bhattacharyya, B.; Doloi, B.N.; Sorkhel, S.K. Experimental investigations into electrochemical discharge machining (ECDM) of non-conductive ceramic materials. *J. Mater. Process. Technol.* **1999**, *95*, 145–154. [CrossRef]
7. Peng, W.Y.; Liao, Y.S. Study of electrochemical discharge machining technology for slicing non-conductive brittle materials. *J. Mater. Process. Tech* **2004**, *149*, 363–369. [CrossRef]
8. Lijo, P.; Somashekhar, S.H. Improvement in Machining Rate with Mixed Electrolyte in ECDM Process. *Procedia Technol.* **2016**, *25*, 1250–1256.
9. Liu, H.; Xieeryazidan, A. Study of Gas Film Characteristics in Electrochemical Discharge Machining and Their Effects on Discharge Energy Distribution. *Micromachines* **2023**, *14*, 1079. [CrossRef]
10. Wang, H. Research on Parameter Optimization of insulator ceramic Tracking electrolytic EDM. *Mod. Manuf. Eng.* **2020**, *482*, 92–98.
11. Cao, Z.H.; Zhang, Z.Y.; Wang, R.; Zhao, D.Y.; Zhu, S.J.; Zhu, H. Characteristics of laser assisted electrochemical Discharge Machining of Microchannels. *Micronanoelectron. Technol.* **2019**, *56*, 480–485.
12. Shi, Z.C.; Jiang, Y.; Ping, X.L.; Zhao, W.S. Application of fuzzy control system in electrochemical discharge wire-cut machining. *Mod. Manuf. Eng.* **2019**, *471*, 80–85.
13. Goud, M.; Sharma, A.K.; Jawalkar, C. A review on material removal mechanism in electrochemical discharge machining (ECDM) and possibilities to enhance the material removal rate. *Precis. Eng.* **2016**, *45*, 1–17. [CrossRef]
14. Jawalkar, C.S.; Sharma, A.K.; Kumar, P. Micromachining with ECDM: Research potentials and experimental investigations. *World Acad. Sci. Eng. Technol.* **2012**, *40*, 7–12.
15. Cheng, C.-P.; Wu, K.L.; Mai, C.-C.; Yang, C.-K.; Hsu, Y.S.; Yan, B.-H. Study of gas film quality in electrochemical discharge machining. *Int. J. Mach. Tools Manuf.* **2010**, *50*, 689–697. [CrossRef]
16. Kumanan, S. A review on current research aspects in tool-based micromachining processes. *Mater. Manuf. Process.* **2014**, *29*, 1291–1337.
17. Singh, M.; Singh, S.; Arora, J.D.; Antil, P.; Oza, A.D.; Burduhos-Nergis, D.D.; Burduhos-Nergis, D.P. Multi Response Optimization of ECDM Process for Generating Micro Holes in CFRP Composite Using TOPSIS Methodology. *Polymers* **2022**, *14*, 5291. [CrossRef]
18. Antil, P.; Singh, S.; Manna, A. Genetic Algorithm Based Optimization of ECDM Process for Polymer Matrix Composite. *Mater. Sci. Forum* **2018**, *928*, 144–149. [CrossRef]
19. Khan, A.A. Electrode wear and material removal rate during EDM of aluminum and mild steel using copper and brass electrodes. *Int. J. Adv. Manuf. Technol.* **2008**, *39*, 482–487. [CrossRef]
20. Sumit, K.J.; Abishek, B.K.; Murali, M.S. High aspect ratio micromachining of glass by electrochemical discharge machining (ECDM). *J. Manuf. Process.* **2013**, *15*, 460–466.
21. Kumar, T.A.; Sekhar, P.S. Optimization of process parameters in ECDM machining using Taguchi based grey relation analysis. *Measurement* **2023**, *216*, 112971.
22. Bian, J.X.; Ma, B.J.; Liu, X.F.; Qi, L.J. Experimental Study of Tool Wear in Electrochemical Discharge Machining. *Appl. Sci.* **2020**, *10*, 5039. [CrossRef]
23. Xie, J.Q. Simulation and experimental study on alumina ceramic EDM. *Harbin Inst. Technol.* **2020**, *6*. [CrossRef]
24. Fukuzawa, Y.; Mohri, N.; Tani, T. Electrical discharge machining phenomena of insulating sialon ceramics with an assisting electrode. *Int. J. Electr. Machin.* **2004**, *2*, 25–30.
25. Mohri, N.; Fukuzawa, Y. Some considerations to machining characteristics of insulating ceramics. *Ann. CIRPVol* **2002**, *14*, 16–30.
26. Fukuzawa, Y.; Iwano, E.; Mohri, N. Micro hole electrical discharge machining on insulating ceramics with pipe electrode. *Int. J. Electr. Machin.* **2003**, *8*, 57–61.
27. Jain, V.K.; Rao, P.S.; Choudhary, S.K.; Rajurkar, K.P. Experimental investigations into traveling wire electrochemical spark machining (TW-ECSM) of composites. *J. Manuf. Sci. Eng.* **1991**, *113*, 75–84.
28. Fascio, V.; Wüthrich, R.; Bleuler, H. Spark assisted chemical engraving in the light of electrochemistry. *Electrochim. Acta* **2004**, *49*, 3997–4003. [CrossRef]
29. Wang, J.; Fu, C.; Jia, Z.X. Cutting of hard and brittle insulating materials using spark discharge-assisted diamond wire sawing. *J. Mater. Process. Tech.* **2018**, *252*, 225–232. [CrossRef]
30. Singh, M.A.; Sarma, D.K.; Hanzel, O.; Sedlacek, J.; Sajgalik, P. Machinability analysis of multi walled carbon nanotubes filled alumina composites in wire electrical discharge machining process. *J. Eur. Ceram. Soc.* **2017**, *37*, 3107–3114. [CrossRef]
31. Moore, M.A.; King, F.S. Abrasive wear of brittle solids. *Wear* **1980**, *60*, 123–140. [CrossRef]

32. Pachaury, Y.; Tandon, P. An overview of electric discharge machining of ceramics and ceramic based composites. *J. Manuf. Process* **2017**, *25*, 369–390. [CrossRef]
33. Roethel, F.; Kosec, L.; Garbajs, V. Contribution to the Micro-Analysis of the Spark Eroded Surfaces. *Ann. CIRP* **1976**, *25*, 135–140.
34. Guu, Y.H.; Hou, M.T.K. Effect of machining parameters on surface textures in EDM of Fe-Mn-Al alloy. *Mater. Sci. Eng. A* **2007**, *466*, 61–67. [CrossRef]
35. Clark, W.I.; Shih, A.J.; Lemaster, R.L.; McSpadden, S.B. Fixed abrasive diamond wire machining—Part II: Experiment design and results. *Int. J. Mach. Tools Manuf.* **2003**, *43*, 533–542. [CrossRef]

applied sciences

Article

Surface Quality Evolution Model and Consistency Control Method of Large Shaft Multi-Pass Grinding

Liping Wang [1,2], Shuailei Fu [1,2], Dong Wang [1,2,*] and Xuekun Li [1,2,*]

1 State Key Laboratory of Tribology, Department of Mechanical Engineering, Tsinghua University, Beijing 100084, China
2 Beijing Key Lab of Precision/Ultra-precision Manufacturing Equipment and Control, Beijing 100084, China
* Correspondence: d-wang@mail.tsinghua.edu.cn (D.W.); xli@tsinghua.edu.cn (X.L.)

Abstract: Large shaft usually achieves high surface quality through multi-pass grinding in practice. Common surface quality indexes include surface roughness and glossiness, which are not only required numerically, but also require high consistency of distribution along the whole shaft. In multi-pass grinding, these two indexes are affected by the process parameters and the surface quality of the previous grinding pass, which leads to the difficulty of modeling. In addition, due to the uneven distribution of actual grinding depth, the surface quality along the whole shaft is usually inconsistent, resulting in the need for multiple spark-out grinding passes to ensure consistency. In this study, the surface quality evolution models for surface roughness and glossiness based on Elman neural network are developed, which build regressions between process parameters, surface quality indexes of the previous grinding pass, and surface quality indexes of the current grinding pass. Moreover, a consistency control method of surface quality is proposed by adjusting the actual grinding depth within the dimensional accuracy tolerance range at the rough grinding stage. Experimental results show that the surface roughness and glossiness prediction errors of the surface quality evolution models are only 5.5% and 5.1%. The consistency control method guarantees the consistency of surface quality, reduces the grinding passes, and increases the grinding efficiency.

Keywords: surface quality; Elman neural network; consistency control method; large shaft grinding; multi-pass grinding

check for updates

Citation: Wang, L.; Fu, S.; Wang, D.; Li, X. Surface Quality Evolution Model and Consistency Control Method of Large Shaft Multi-Pass Grinding. *Appl. Sci.* **2023**, *13*, 1502. https://doi.org/10.3390/app13031502

Academic Editors: Rui Manuel Leal, Gustavo H. S. F. L. Carvalho and Ivan Galvão

Received: 11 January 2023
Revised: 18 January 2023
Accepted: 21 January 2023
Published: 23 January 2023

1. Introduction

Large shaft is widely used in industry to bear heavy load and transfer motion [1], and multi-pass grinding is the final process to obtain high surface quality [2]. Surface roughness and glossiness are common surface quality indexes, and the requirements of these two indexes are high in large shaft grinding. Moreover, surface quality consistency needs to be ensured, which means the value difference of each surface quality index along the whole shaft should be small. Both surface roughness and glossiness are affected by grinding parameters, so the reasonable selection of grinding parameters is the basis to obtain the desired surface quality [3,4].

Some studies reveal the mechanism of grinding surface formation through physical modeling of the grinding process, and then obtain the correlation between grinding parameters and surface roughness [5,6]. Zhou et al. proposed a conventional method determining the surface roughness using the mean value of the grain protrusion heights [7]. Jiang et al. developed a predictive model for grinding surface roughness based on micro-interacting mechanism modeling [8]. Chakrabart et al. simplified the abrasive particles into a cone and calculated the surface roughness by obtaining the topography of the grinding surface through a simulation method [9]. The above research established surface roughness prediction models based on physical modeling and simulation, but these models are difficult to describe the random factors in the actual grinding process, so the prediction performances are not high enough. Another kind of surface roughness modeling method is the empirical method, by which the correlation between grinding parameters and surface roughness is established by numerical method [10,11]. Kwak et al. proposed a surface roughness prediction model based on a respond surface model for external cylindrical grinding of SCM440 steel [12]. Alao et al. used the Taguchi method to establish the correlation between grinding depth, grinding wheel speed, and surface roughness [13]. Caydas et al. developed the least squares vector machine and spider support vector machine to predict the surface roughness of stainless steel turning [14]. Liu et al. used an artificial neural network to accurately describe the influence of grinding parameters on surface roughness [15]. In general, the empirical method is the most commonly used method for predicting surface roughness with grinding parameters due to its convenient calculation and high precision [16,17]. Recently, some studies predict surface roughness by monitoring grinding signals, such as image [18] and acoustic emission signals [19]. However, this method is not helpful to the selection of grinding parameters.

Glossiness is an optical parameter to represent the reflective capacity of the surface, which also reflects the surface roughness [20]. Generally, the higher the surface roughness, the lower the glossiness [21]. The existing research seldom discusses the correlation between process parameters and glossiness based on the modeling of the machining process, but mostly focuses on the correlation between roughness and glossiness. Yavuz et al. performed polishing experiments on building stone tiles with SiC abrasive particles and found an exponential relationship between the surface roughness and glossiness of the workpiece [22]. Nevertheless, by testing polymer surfaces, Assender et al. pointed out that surfaces with similar roughness can also have very different glossiness [23]. Li et al. simulated the surface roughness and glossiness with non-Gaussian topography features, and it is found that there can be other indexes deciding the final glossiness besides surface roughness [24]. Overall, current research on glossiness modeling is insufficient.

The above researchers have made achievements in the building correlation between grinding surface quality and process parameters, but there are still two limitations. Firstly, the influence of the surface quality of the previous grinding pass is not considered by existing research. In multi-pass grinding, the same grinding parameters will result in different grinding qualities under different initial surface qualities, which makes the surface quality model ineffective. Secondly, the research on the formation mechanism and control method of surface quality consistency is insufficient. Even if better grinding parameters are selected to obtain higher grinding quality, there will still be large differences in the grinding quality indexes along the whole shaft; in other words, the grinding quality consistency is poor. Thus, the surface quality consistency is hard to guarantee through selection of grinding parameters alone.

In this study, the formation mechanism of surface quality consistency is revealed by qualitative analysis of the influence of grinding parameters on surface quality. Then, the surface quality evolution models for surface roughness and glossiness based on Elman neural network are developed, which build regressions between process parameters, surface quality indexes of the previous grinding pass, and surface quality indexes of the current grinding pass. Moreover, a consistency control method of surface quality is proposed by adjusting the actual grinding depth within the dimensional accuracy tolerance range at the rough grinding stage. A case study is performed to verify the effectiveness of the proposed method.

2. Surface Quality Indexes and Influencing Factors

2.1. Surface Quality Indexes

The surface quality indexes in this study are surface roughness and glossiness. Surface roughness is used to measure surface fluctuation. In this study, the arithmetic mean deviation Ra is selected as the surface roughness value. Glossiness is a common index used to evaluate finished surfaces, such as roller grinding and parts polishing [24]. Glossiness G is expressed as:

$$G = \frac{R}{R_0} \times 100\% \tag{1}$$

where R is the specular reflectance of the measured surface, and R_0 is the specular reflectance of the standard glass, which is the polished black glass with a refractive index of 1.567. The test method for glossiness is shown in Figure 1. The light source illuminates the measured surface at a specific incident angle and is reflected at an angle of the same magnitude as the incident angle. The reflected light is received by the detector, and the ratio of the reflected light energy to the incident light energy is calculated as the specular reflectance of the measured surface R. Mirror reflection of the standard glass is performed to obtain R_0, and the glossiness of the measured surface can be calculated according to Equation (1). It should be noted that the test value of glossiness changes when the incident angle of light is changed. The ASTM D523 standard test method developed by Hunter and ASTM (American Society for Testing and Materials) specifies the test of glossiness at 60° incident light, so the angle of incidence used in this paper is 60°.

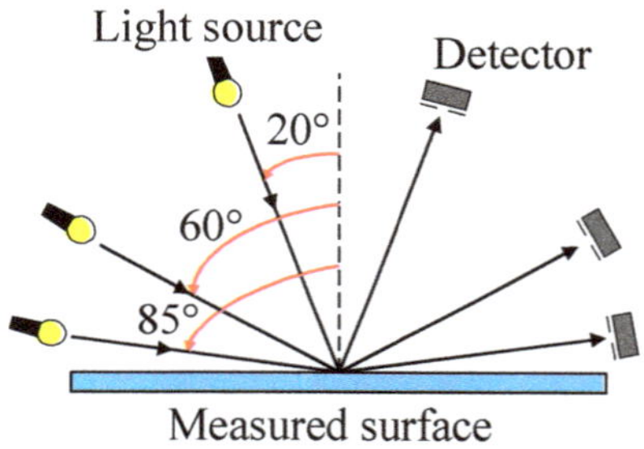

Figure 1. The test method for glossiness.

2.2. Analysis of Surface Quality Influencing Factors

The full factor experiments are carried out to analyze the influence of grinding parameters on surface roughness and glossiness. The grinding machine is a grinder MKT8445. The workpiece is a shaft with a length of 1480 mm and a diameter of 244 mm. The material of the shaft is 92CrMo, which is a kind of alloy steel. The grinding wheel is made of silicon carbide with a width of 60 mm. The grinding system and process are shown in Figure 2a,b. The shaft diameter is obtained by measuring arm as shown in Figure 2c, and the measuring resolution is 0.1 μm. The surface roughness is tested by Surftest-SJ210 roughness meter with a resolution of 0.001 μm. The direction of measurement of roughness is parallel to the direction of grinding passes as shown in Figure 2d. The glossiness is tested by the YG60 glossiness meter with a resolution of 1GU and a test angle of 60°. The testing instruments are shown in Figure 2e,f.

The full factor experiments consist of four factors, i.e., grinding depth ap, grinding wheel speed Nwheel, shaft speed Nshaft, and pallet speed f. The grinding depth and the grinding wheel speed are set at four levels, while the shaft speed and the pallet speed are set at three levels as shown in Table 1. Thus, the full factor experiments consist of 144 grinding passes.

Table 1. Grinding parameters of the full factor experiments.

Level	a_p (μm)	N_{wheel} (r/min)	N_{shaft} (r/min)	f (mm/min)
1	0	550	60	800
2	2	700	70	1000
3	3.5	850	80	1200
4	5	1000		

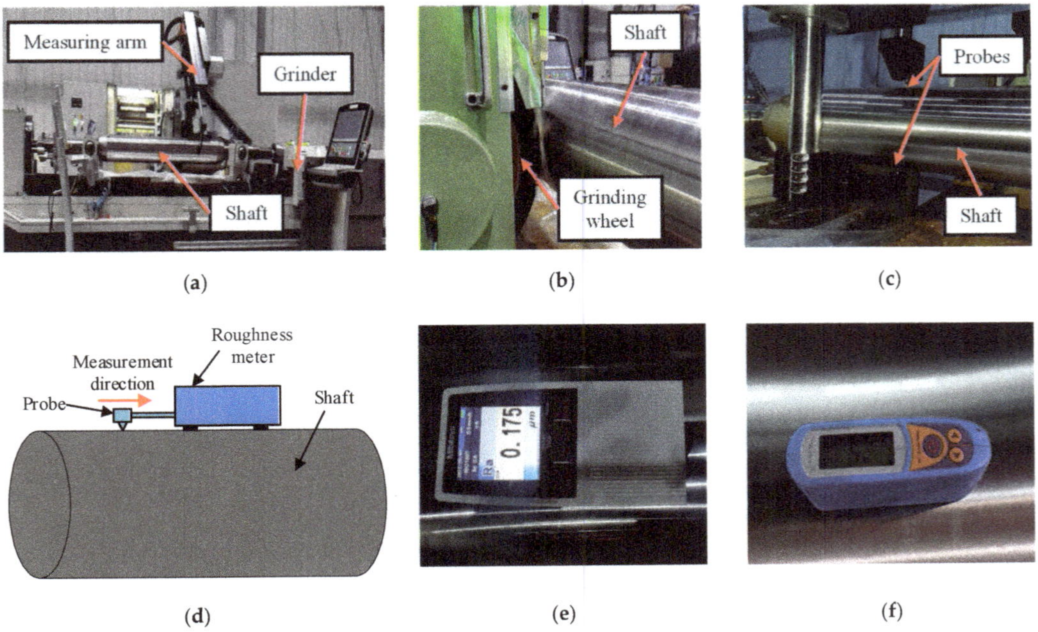

Figure 2. The grinding system and testing instruments: (**a**) grinding system, (**b**) grinding process, (**c**) diameter measurement, (**d**) roughness measurement, (**e**) roughness meter, (**f**) glossiness meter.

The grinding wheel grinds from the left side of the shaft to the right side in a grinding pass. On the next grinding pass, the grinding wheel grinds from the right side of the shaft to the left side. In order to expand the number of samples, six measuring points are selected along the z-direction on the shaft after grinding to test the surface roughness and glossiness as shown in Figure 3. The measuring points in this study are distributed uniformly along the length of the shaft, which can reflect the trend of surface quality on the whole shaft.

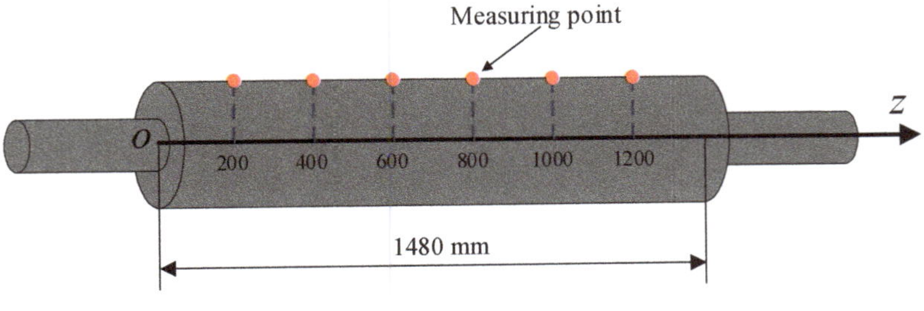

Figure 3. Measuring points of shaft.

Figure 4 shows the surface roughness and glossiness with different grinding parameters, and the values of surface roughness and glossiness are average values tested on 6 measuring points. The error bar represents the standard deviation. With the increase of grinding depth, shaft speed, and pallet speed, the surface roughness value increases and the glossiness decreases. This is because the increase of these three grinding parameters will increase the grinding force, resulting in more intense plastic deformation of the shaft and a rougher surface.

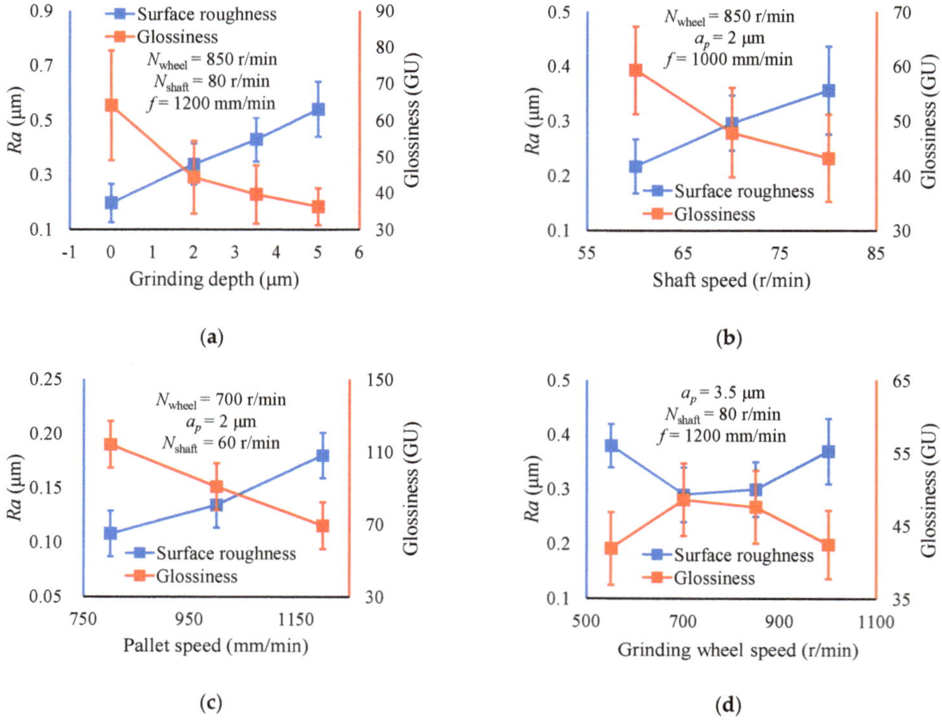

Figure 4. The influence of three grinding parameters on surface roughness and glossiness: (**a**) grinding depth, (**b**) shaft speed, (**c**) pallet speed, (**d**) grinding wheel speed.

For grinding wheel speed, lower surface roughness values and higher glossiness values are obtained in the range from 700 to 850 r/min. The spindle current standard deviations of the grinding passes in Figure 5d are shown in Table 2. It can be seen that the current variances are small when the grinding wheel speeds are 700 r/min and 850 r/min. The spindle runs more stably within this speed range, and the contact between the grinding wheel and the shaft is more stable, resulting in higher surface quality.

Table 2. The spindle current standard deviations of the grinding passes in Figure 5d.

N_{wheel} (r/min)	550	700	850	1000
Standard deviations (A)	1.35	0.64	0.73	1.23

It should be noted that the grinding depth involved in the above analysis is the nominal grinding depth. In fact, due to grinding wheel wear in the grinding process, the actual grinding depth is less than the nominal grinding depth and unevenly distributed along the *z*-direction. Therefore, it is necessary to further analyze the actual grinding depth at different positions of the same grinding pass and its influence on the surface roughness and glossiness. The actual grinding depth a'_p is calculated as:

$$a'_p = \frac{1}{2}\left(D_{\text{shaft}} - D'_{\text{shaft}}\right) \tag{2}$$

where D_{shaft} and D'_{shaft} are diameters of the shaft before and after a grinding pass.

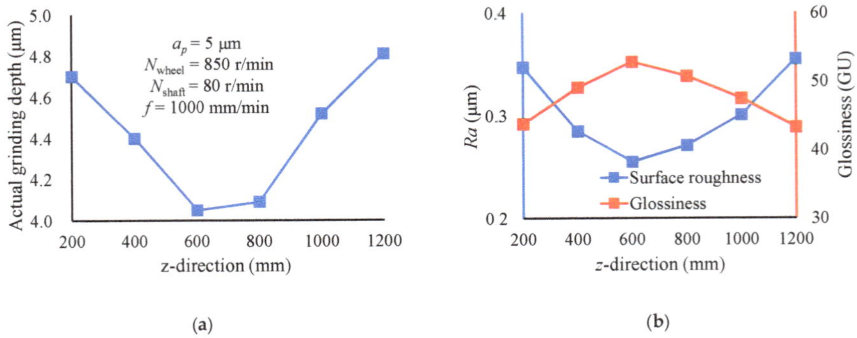

Figure 5. The actual grinding depth, surface roughness, and glossiness of a single grinding pass: (**a**) actual grinding depth, (**b**) surface roughness and glossiness.

Figure 5 shows the actual grinding depth, surface roughness, and glossiness of a single grinding pass. The actual grinding depth is larger on both sides of the shaft and smaller in the middle due to grinding wheel wear. With the decrease of actual grinding depth, the roughness value decreases and the glossiness increases, which is consistent with the effect of nominal grinding depth on surface quality. This reveals the mechanism of inconsistent surface quality. The actual grinding depth along the z-direction is inconsistent due to grinding wheel wear, so that the surface roughness and glossiness are inconsistent along the z-direction. In addition, the deflection of the shaft and grinding wheel under grinding force will also reduce the actual grinding depth. However, in this study, the shaft size is large and the grinding depth is small (less than 5 μm), so the influence of grinding force can be ignored.

The grinding process of large shaft parts usually includes four stages: rough grinding, semi-finish grinding, finish grinding, and spark-out grinding. Each stage includes several grinding passes with the same parameters. In a grinding stage, even if the grinding parameters are the same, the surface roughness and glossiness after grinding are different. Figure 6 shows the surface roughness and glossiness of four consecutive passes with the same grinding parameters. The data in Figure 6 are the test values only at the measuring point at z = 800 mm. As can be seen from Figure 6, these two surface quality indicators change with the grinding passes. This indicates that the surface quality of large shaft parts in multi-pass grinding is affected by the quality of the previous grinding pass, showing evolutionary characteristics, rather than a simple correlation only with grinding parameters. It should be pointed out that the above conclusions are obtained when the material of workpiece is 92CrMo alloy steel. Considering that the material of the large shaft is essentially alloy steel, this conclusion applies to other large shafts made of different kinds of alloy steel.

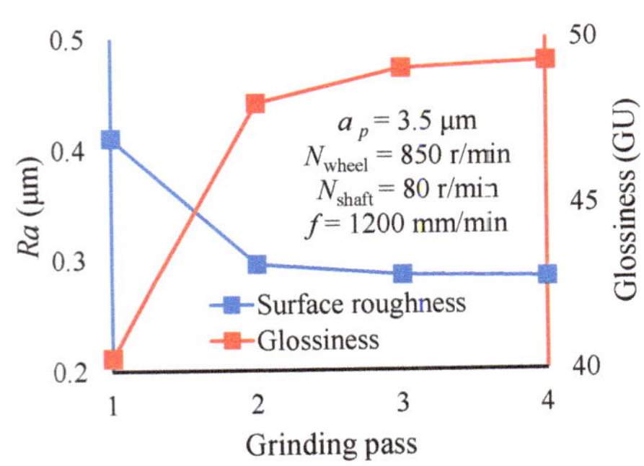

Figure 6. The surface roughness and glossiness of four consecutive passes.

3. Surface Quality Evolution Model Based on Elman Neural Network

The analysis in Section 2.2 reveals that the surface quality has evolutionary characteristics. Therefore, the surface quality evolution model is established based on Elman neural network, which is suitable to process data with evolutionary characteristics. Elman neural network consists of input layer, hidden layer, undertake layer, and output layer, and its architecture is shown in Figure 7.

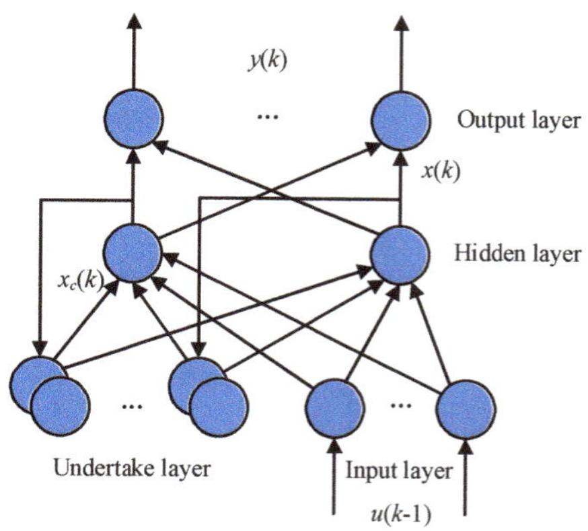

Figure 7. Architecture of Elman neural network.

Assuming that the input of the network is $u(k-1)$, the output is $y(k)$, the input of the hidden layer is $x(k)$, and the output of the undertake layer is $x_c(k)$, the state space of the Elman neural network can be expressed as:

$$x(k) = f(w_k^1 x_c(k) + w_k^2 u(k-1)) \tag{3}$$

$$x_c(k) = x(k-1) \tag{4}$$

$$y(k) = g(w_k^3 x(k)) \tag{5}$$

where w_k^1, w_k^2 and w_k^3 are the weight matrices of the undertake layer, input layer, and hidden layer, respectively; $f(\cdot)$ and $g(\cdot)$ are the transfer functions of the input layer and the hidden layer, which are tansig function and purelin function, respectively. As can be seen from Equations (2)–(4), the input of the network includes not only the input of the input layer, but also the previous output value of the hidden layer, which makes the network sensitive to the historical state data and suitable to process data with evolutionary characteristics.

Based on the above analysis, the surface quality evolution model based on Elman neural network is developed. The models for surface roughness and glossiness are established respectively. For the model of surface roughness, Elman neural network is trained by the data from the full factor experiments. The input layer of the network has five nodes, which respectively correspond to the four grinding parameters of the current grinding pass, i.e., the actual grinding depth, grinding wheel speed, shaft speed and pallet speed, and the surface roughness of the previous grinding pass. The output of the network has one node that corresponds to the surface roughness of the current pass. The model of glossiness is similar to the model of surface roughness, except that the surface quality index of input and output is glossiness.

The surface quality evolution models for surface roughness and glossiness are trained using the data from the full factor experiment. The full factor experiments consist of 144 grinding passes, each of which contained six measuring points, resulting in a total of 864 samples. 700 samples are used for training and 164 samples for testing. After the training of 20,000 epochs, the prediction errors of surface roughness and glossiness reach

5.5% and 5.1%, respectively. The actual values of surface roughness and glossiness and predicted values from surface quality evolution models are shown in Figure 8.

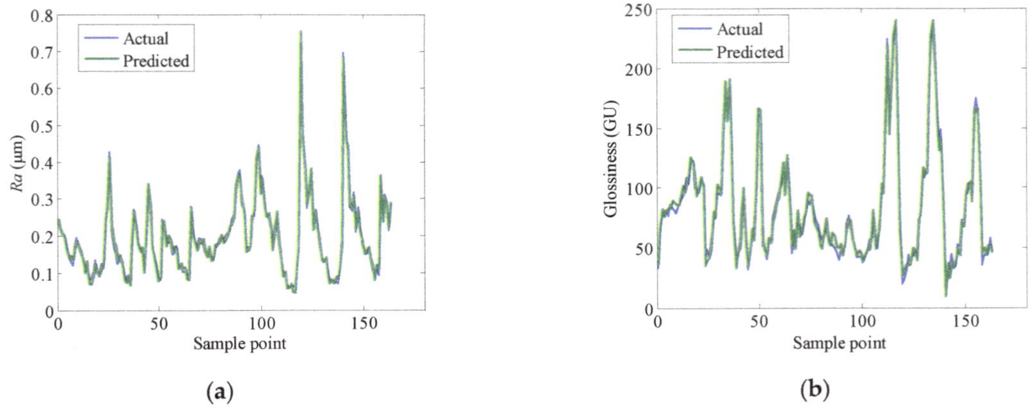

Figure 8. Prediction performance of surface quality evolution models: (**a**) surface roughness, (**b**) glossiness.

In order to verify the superiority of the proposed model, the prediction accuracy of different modeling methods for surface roughness and glossiness is compared. Elman neural network is used in Models 1 to 3. The training samples and parameters are consistent with the proposed model, but there are some differences in the input and output of the model. Model 1 uses nominal grinding depth. Model 2 ignores the influence of the previous grinding surface quality; Model 3 uses nominal grinding depth and ignores the effect of the previous grinding surface quality. The prediction performance of these three models is shown in Table 3.

Table 3. The prediction performance of different models for surface quality indexes.

Models	Grinding Depth	Previous Grinding Surface Quality	Average Prediction Error of Surface Roughness	Average Prediction Error of Glossiness
Proposed model	Actual	Considered	5.5%	5.1%
Model 1	Nominal	Considered	12.4%	13.2%
Model 2	Actual	Ignored	15.3%	16.8%
Model 3	Nominal	Ignored	24.2%	25.3%

It can be seen from Table 3 that the prediction errors of the proposed model for surface roughness and glossiness are minimal. This proves that the grinding surface quality index is affected by the actual grinding depth and the quality of the previous grinding, and also proves the superiority of the proposed model.

4. Consistency Control Method of Surface Quality in Rough Machining

4.1. The Principle of Consistency Control Method of Surface Quality

Since the surface quality is affected by actual grinding depth, a consistency control method of surface quality is proposed based on adjustment of actual grinding depth. Considering the grinding depth is quite small in practice, the adjustment of actual grinding depth is only performed in rough grinding.

The principle of the adjustment of actual grinding depth is shown in Figure 9. The profile of the shaft before a grinding pass and the ideal profile after a grinding pass are shown in Figure 9a. The blue area is the tolerance range. The maximum and minimum values of the actual grinding depth allowed at z_0 are $\max(a_p(z_0))$ and $\min(a_p(z_0))$, respectively.

Therefore, the allowable range of the actual grinding depth along the z-direction can be obtained, as shown in Figure 9b.

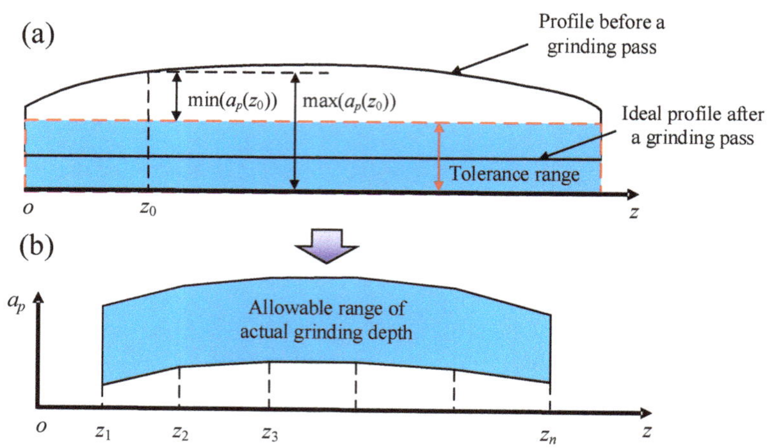

Figure 9. The principle of the adjustment of actual grinding depth: (**a**) the profile of the shaft before a grinding pass and the ideal profile after a grinding pass, (**b**) the allowable range of the actual grinding depth.

Several points z_1, z_2...z_n are selected along the z-direction. According to the actual grinding depth range allowed at each point and the surface quality evolution model established in Section 4, the adjustable region of surface roughness and glossiness can be obtained as shown in Figure 10. The maximum values and minimum values of surface roughness and glossiness of the adjustable region are Ra_0, Ra_1, G_0, and G_1.

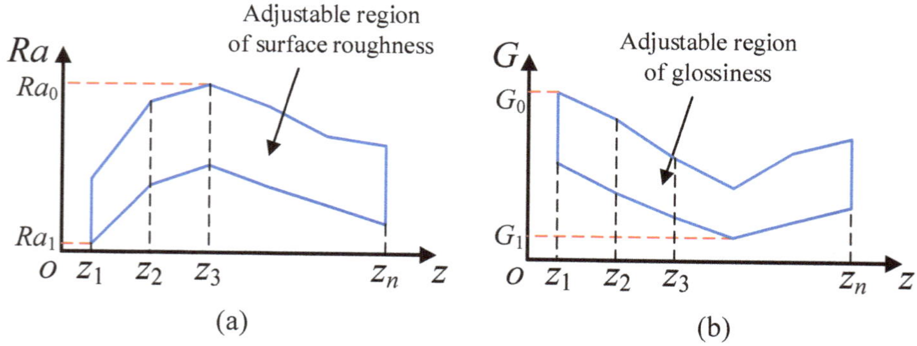

Figure 10. The adjustable region of (**a**) surface roughness and (**b**) glossiness.

The comprehensive index of surface quality, C, is defined as:

$$C = \frac{Ra_0 - Ra}{2(Ra_0 - Ra_1)} + \frac{G - G_1}{2(G_0 - G_1)} \quad (6)$$

The adjustable range of the comprehensive index of surface quality will have two situations, as shown in Figure 11. In Figure 11a, the minimum value of the upper bound of the adjustable region is completely contained within the adjustable region. Therefore, C of each point can reach C_0 by adjusting the actual grinding depth. In Figure 11b, the minimum value of the upper bound of the adjustable region is not fully contained within the adjustable region. Then, the points that can obtain C_0 are adjusted to obtain C_0, and the points that cannot obtain C_0 are adjusted to the value closest to C_0.

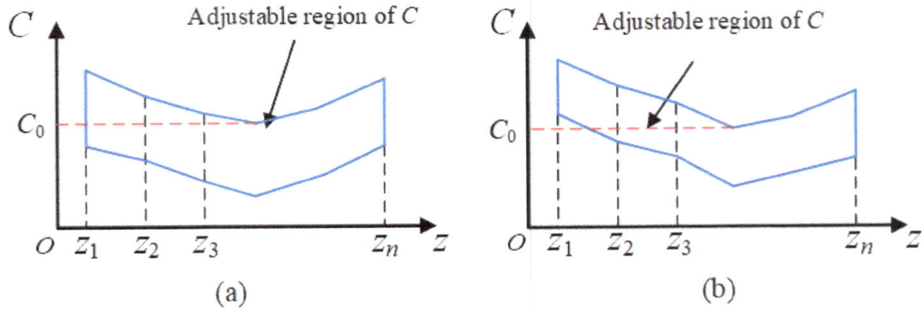

Figure 11. The adjustable region of C: (**a**) the adjustable region of C for situation 1, (**b**) the adjustable region of C for situation 2.

The process of obtaining the desired comprehensive index C_i for point z_i by grinding depth adjustment is shown in Figure 12.

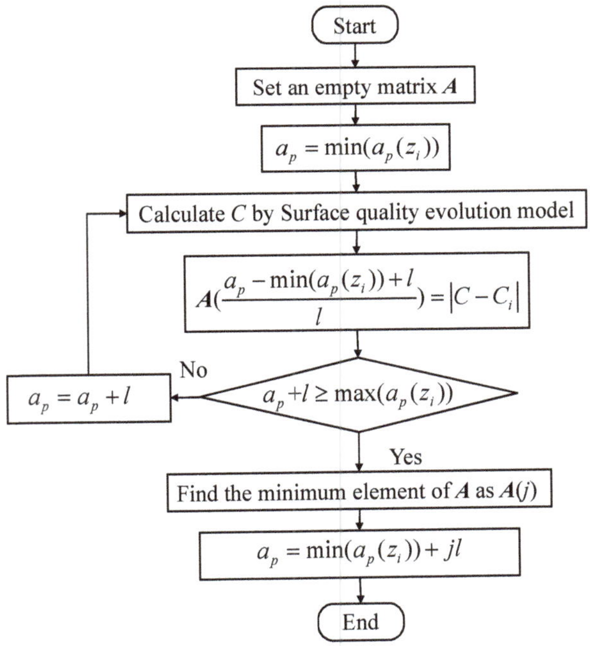

Figure 12. The process of obtaining the desired comprehensive index C_i for point z_i.

In Figure 12, l is the iteration step length. After obtaining the actual grinding depth values of each point, the actual grinding depth curve along the z-direction is generated by B-spline interpolation.

4.2. The Compensation Method to Guarantee the Actual Grinding Depth

Our previous research indicated that grinding wheel wear is the main factor causing grinding dimensional error of large shaft parts, and the dimensional error model based on grinding wheel wear and compensation method was proposed [2]. Therefore, dimensional errors can be predicted based on grinding process parameters and shaft profiles. Based on the predicted dimensional error, the grinding wheel path can be adjusted to guarantee the actual grinding depth. Experiments show that the actual grinding depth error of the compensation method in reference [2] is only 0.16 μm for the shaft when the nominal grinding depth is 5 μm in this study as shown in Figure 13. It is proved that the compensation method is effective and the main factor causing grinding dimensional error is grinding wheel wear.

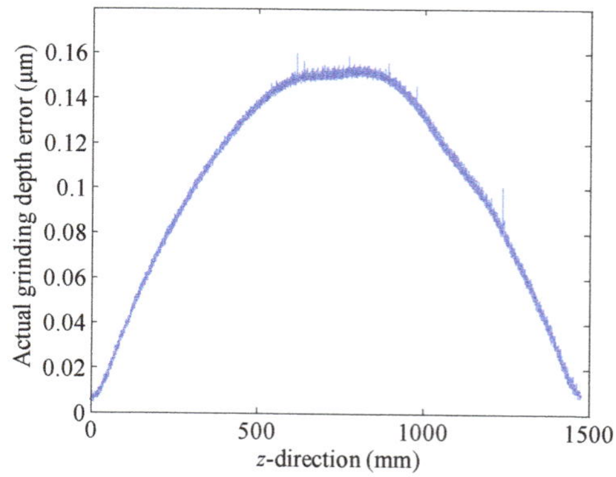

Figure 13. Actual grinding depth error after compensation.

5. Case Study

5.1. Experimental Set Up

A case study is conducted in this section to verify the proposed methods, i.e., the surface quality evolution model and a consistency control method of surface quality. The grinding machine, grinding wheel, workpiece, and testing instruments used in the experiment are the same as those described in Section 2.

In case study, the complete grinding process is carried out, including rough grinding, semi-finish grinding, finish grinding, and spark-out grinding. The parameters of these four grinding stages are shown in Table 4.

Table 4. The parameters of grinding process.

Grinding Stage	a_p (μm)	N_{wheel} (r/min)	N_{shaft} (r/min)	f (mm/min)	Grinding Pass
Rough	5	1000	80	1200	3
Semi-finish	3	900	70	1050	2
Finish	2	800	70	1000	2
Spark-out	0	800	60	1000	3

Six measuring points are selected on the axis with z-coordinates of 200, 400, 600, 800, 1000, and 1200 mm, respectively. Before each grinding pass in the rough grinding, the profile of the shaft along the z-direction is measured and the actual grinding depth is adjusted by the consistency control method. The dimensional error is predicted according to the process parameters, and the grinding wheel pass is adjusted by compensation method to guarantee the actual grinding depth. The iteration step length l is 0.02 μm, and the tolerance range of machining error is 1 μm in rough grinding. In the semi-finish grinding and finish grinding, only the dimensional error is compensated, while the consistency of surface quality is not controlled. In spark-out grinding, the grinding wheel path is not adjusted, since the grinding depth is 0. After each grinding pass, the surface roughness and glossiness of the measuring points are tested.

5.2. Results and Discussion

The standard deviations of surface roughness and glossiness of measuring points are used to describe the surface quality consistency as shown in Figure 14, and it can be seen that the surface quality consistency gradually becomes better during the grinding process. The standard deviations of surface roughness and glossiness of measuring points are 0.651 μm and 26.2 GU for the first grinding pass, and 0.060 μm and 4.1 GU for the last grinding pass, respectively.

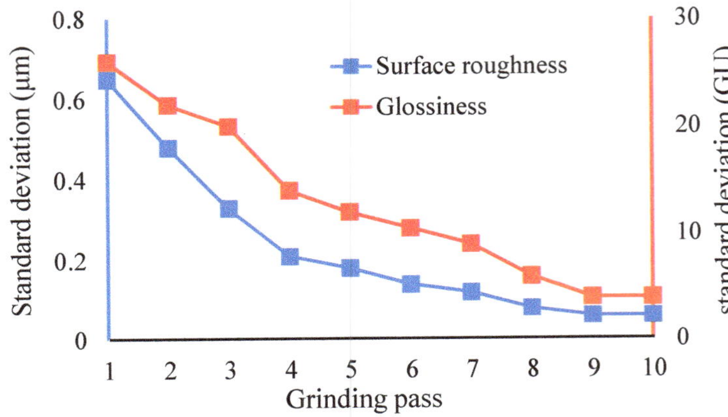

Figure 14. The standard deviations of surface roughness and glossiness of measuring points.

The dimensional errors of rough grinding passes are shown in Figure 15. It can be seen that the dimensional errors are still within the tolerance range after the adjustment of the actual grinding depth, which proves the effectiveness of the compensation method. It is also proved that the consistency control method of surface quality will not make the dimensional error exceed the tolerance range.

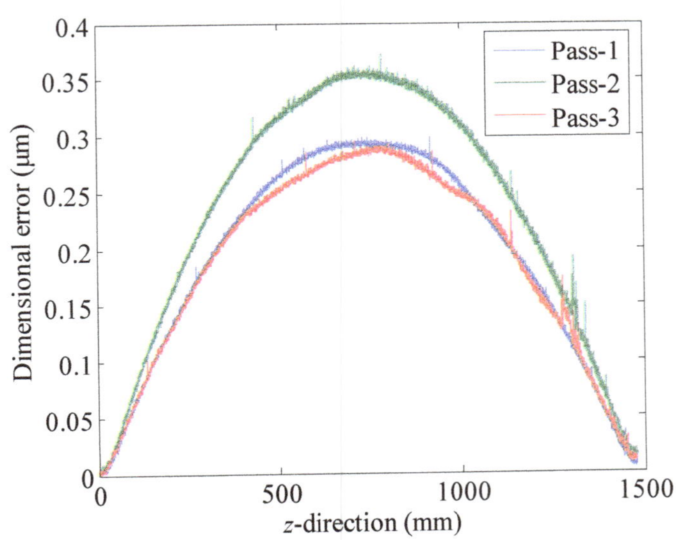

Figure 15. The dimensional errors of rough grinding passes.

The shaft is continued to be machined with rough grinding process parameters, without compensation and the consistency control method. The standard deviations of surface roughness and glossiness of measuring points become 0.628 μm and 23.1 GU. Then, the shaft is machined again with process parameters in Table 4. During rough grinding, semi-finish grinding, and finish grinding, only the dimensional error is compensated, while the consistency of surface quality is not controlled. Finally, enough spark-out grinding passes are performed until the standard deviations of surface roughness and glossiness are close to the standard deviations of the last grinding pass in Figure 14. After eight spark-out grinding passes, the standard deviations of surface roughness and glossiness are 0.055 μm and 4.1 GU.

The standard deviations of surface roughness and glossiness are shown in Figure 16. It can be seen that without the consistency control method, it requires five more spark-out grinding passes to achieve high consistency. This shows that the consistency control method of surface quality can obtain a higher consistency of surface quality with reducing the grinding passes, resulting in high grinding efficiency.

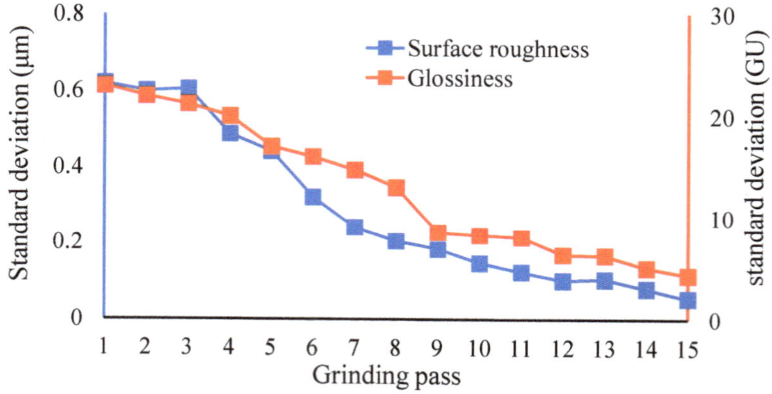

Figure 16. The standard deviations of surface roughness and glossiness of measuring points.

6. Conclusions

This research studies the surface quality evolution and consistency control of large shaft multi-pass grinding. Based on the analysis of the grinding experimental data, it is found that the grinding surface quality is affected by the grinding process parameters and the surface quality of the previous grinding pass, which shows the evolutionary characteristics. Then, the surface quality evolution model is established based on Elman neural network. Furthermore, the consistency control method of surface quality is proposed by adjusting the actual grinding depth at the rough grinding stage. Experiments show the effectiveness of the proposed methods. The conclusions drawn from this research can be summarized as follows:

(1) The surface quality indexes of multi-pass grinding of shaft parts, i.e., surface roughness and glossiness, have evolutionary characteristics and are affected by grinding parameters and surface quality of the previous grinding pass. The uneven distribution of the actual grinding depth along the length direction of the shaft is the reason for the inconsistent surface quality.

(2) The surface quality evolution model based on Elman neural network accurately describes the evolution characteristics of surface quality, and the prediction errors of surface roughness and glossiness are less than 6%.

(3) The consistency control method of surface quality guarantees the consistency of surface quality, reduces the grinding passes, and increases the grinding efficiency.

Author Contributions: Conceptualization, L.W. and S.F.; methodology, L.W. and S.F.; validation, S.F.; formal analysis, L.W. and S.F.; investigation, S.F.; data curation, L.W.; writing—original draft preparation, L.W. and S.F.; writing—review and editing, D.W. and X.L.; supervision, L.W.; funding acquisition, L.W., D.W. and X.L. All authors have read and agreed to the published version of the manuscript.

Funding: This work is supported by National Natural Science Foundation of China under Grant No.52105520, No.52275440 and No.51975319.

Institutional Review Board Statement: Not applicable.

Informed Consent Statement: Not applicable.

Data Availability Statement: All data included in this study are available upon request by contact with the corresponding authors.

Conflicts of Interest: The authors declare no conflict of interest.

References

1. Chybowski, L.; Nozdrzykowski, K.; Grzadziel, Z.; Jakubowski, A.; Przetakiewicz, W. Method to increase the accuracy of large crankshaft geometry measurements using counterweights to minimize elastic deformations. *Appl. Sci.* **2020**, *10*, 4722. [CrossRef]
2. Wang, C.; Wang, D.; Wang, L.P.; Jiang, S.; Li, H.Y.; Li, X.K. The development of time-dependent compensation model for roller CVC profile generation in precision grinding. *Int. J. Adv. Manuf. Technol.* **2021**, *114*, 1671–1684. [CrossRef]
3. Patel, D.K.; Goyal, D.; Pabla, B.S. Optimization of parameters in cylindrical and surface grinding for improved surface finish. *R. Soc. Open Sci.* **2018**, *5*, 171906. [CrossRef] [PubMed]
4. Zhang, S.Y.; Zhang, G.B.; Ran, Y.; Wang, Z.C.; Wang, W. Multi-objective optimization for grinding parameters of 20CrMnTiH gear with ceramic microcrystalline corundum. *Materials* **2019**, *12*, 1352. [CrossRef] [PubMed]
5. Wang, Y.Z.; Liu, Y.; Chu, X.M.; He, Y.M.; Zhang, W. Calculation model for surface roughness of face gears by disc wheel grinding. *Int. J. Mach. Tool. Manu.* **2017**, *123*, 76–88. [CrossRef]
6. Wang, Y.Z.; Chen, Y.Y.; Zhou, G.M.; Lv, Q.J.; Zhang, Z.Z.; Tang, W.; Liu, Y. Roughness model for tooth surfaces of spiral bevel gears under grinding. *Mech. Mach. Theory.* **2016**, *104*, 17–30. [CrossRef]
7. Zhou, X.; Xi, F. Modeling and predicting surface roughness of the grinding process. *Int. J. Mach. Tool. Manu.* **2022**, *42*, 969–977. [CrossRef]
8. Jiang, J.L.; Ge, P.Q.; Hong, J. Study on micro-interacting mechanism modeling in grinding process and ground surface roughness prediction. *Int. J. Adv. Manuf. Technol.* **2013**, *67*, 1035–1052. [CrossRef]
9. Chakrabarti, S.; Paul, S. Numerical modelling of surface topography in superabrasive grinding. *Int. J. Adv. Manuf. Technol.* **2008**, *39*, 29–38. [CrossRef]
10. Yu, H.D.; Wang, M.X.; Xu, J.K.; Chen, G.J.; Dai, B.; Wang, S. Research on surface roughness prediction model of ultrasonic assisted grinding by response surface method. *Solid State Phenom.* **2021**, *324*, 66–71. [CrossRef]
11. Zhu, C.M.; Gu, P.; Wu, Y.Y.; Liu, D.H.; Wang, X.K. Surface roughness prediction model of SiCp/Al composite in grinding. *Int. J. Mech. Sci.* **2019**, *155*, 98–109. [CrossRef]
12. Kwak, J.S.; Sim, S.B.; Jeong, Y.D. An analysis of grinding power and surface roughness in external cylindrical grinding of hardened SCM440 steel using the response surface method. *Int. J. Mach. Tool. Manu.* **2006**, *46*, 304–312. [CrossRef]
13. Alao, A.R.; Konneh, M. Application of Taguchi and Box-Behnken designs for surface roughness in precision grinding of silicon. *Int. J. Precis. Technol.* **2011**, *2*, 21–38. [CrossRef]
14. Caydas, U.; Ekici, S. Support vector machines models for surface roughness prediction in CNC turning of AISI 304 austenitic stainless steel. *J. Intell. Manuf.* **2012**, *23*, 639–650. [CrossRef]
15. Liu, Y.; Song, S.Y.; Zhang, Y.D.; Li, W.; Xiao, G.J. Prediction of surface roughness of abrasive belt brinding of superalloy material based on RLSOM-RBF. *Materials* **2022**, *14*, 5701. [CrossRef] [PubMed]
16. Lipinski, D.; Balasz, B.; Rypina, L. Modelling of surface roughness and grinding forces using artificial neural networks with assessment of the ability to data generalization. *Int. J. Adv. Manuf. Technol.* **2018**, *94*, 1335–1347. [CrossRef]
17. Zhang, Y.W.; Li, B.Z.; Yang, J.G.; Liang, S. Modeling and optimization of alloy steel 20CrMnTi grinding process parameters based on experiment investigation. *Int. J. Adv. Manuf. Technol.* **2018**, *95*, 1859–1873. [CrossRef]
18. Zhang, G.J.; Liu, C.Y.; Min, K.; Liu, H.; Ni, F.L. A GAN-BPNN-based surface roughness measurement method for robotic grinding. *Machines* **2022**, *10*, 1026. [CrossRef]
19. Yin, G.Q.; Wang, J.H.; Guan, Y.Y.; Wang, D.; Sun, Y. The prediction model and experimental research of grinding surface roughness based on AE signal. *Int. J. Adv. Manuf. Technol.* **2022**, *120*, 6693–6705. [CrossRef]
20. Yonehara, M.; Matsui, T.; Kihara, K.; Isono, H.; Kijima, A.; Sugibayashi, T. Evaluation method of surface texture by surface roughness based on geometrical product specifications (GPS). *Mater. Trans.* **2004**, *45*, 1019–1026. [CrossRef]
21. Son, J.; Lee, H. Preliminary study on polishing SLA 3D-printed ABS-like resins for surface roughness and glossiness reduction. *Micromachines* **2020**, *11*, 843. [CrossRef] [PubMed]

22. Yavus, H.; Ozkahraman, T.; Demirdag, S. Polishing experiments on surface quality of building stone tiles. *Constr. Bulid. Mater.* **2011**, *25*, 1707–1711. [CrossRef]
23. Assender, H.; Bliznyuk, V.; Porfyrakis, K. How surface topography relates to materials properties. *Science* **2002**, *297*, 973–976. [CrossRef] [PubMed]
24. Li, H.Y.; Li, X.K.; Tian, C.C.; Yang, X.D.; Zhang, Y.; Liu, X.L.; Rong, Y.M. The simulation and experimental study of glossiness formation in belt sanding and polishing processes. *Int. J. Adv. Manuf. Technol.* **2017**, *90*, 199–209. [CrossRef]

applied sciences

MDPI

Article

Multi-Objective Optimization of Micro-Milling Parameters—The Trade-Offs between Machining Quality, Efficiency, and Sustainability in the Fabrication of Thin-Walled Microstructures

Peng Wang [1], Qingshun Bai [1,*], Kai Cheng [2], Liang Zhao [1] and Yabo Zhang [1]

[1] School of Mechanical and Electrical Engineering, Harbin Institute of Technology, Harbin 150001, China; zhaoliang309@hit.edu.cn (L.Z.); 19b908121@stu.hit.edu.cn (Y.Z.)

[2] Department of Mechanical and Aerospace Engineering, Brunel University London, Uxbridge UB8 3PH, UK; kai.cheng@brunel.ac.uk

* Correspondence: qshbai@hit.edu.cn

Abstract: Micro-milling has found extensive applications in machining components with thin-walled microstructures, such as terahertz slow-wave structures, microfluidic chips, and micro-molds. Due to the influence of size effects, micro-milling exhibits higher specific energy consumption compared with traditional milling, implying that more energy is consumed to remove a unit volume of material, particularly in challenging-to-machine materials like Ti-6Al-4V. Historically, research on parameter optimization for micro-milling has predominantly focused on enhancing machining quality and efficiency, with limited attention given to energy efficiency. However, in the context of the "double carbon" strategy, energy conservation and emissions reduction have garnered significant attention in the manufacturing industry. Therefore, this paper proposes a micro-milling parameter-based power consumption model. Based on this, a specific energy consumption model can be obtained. Moreover, evolutionary algorithms are utilized for the optimization of micro-milling parameters, which aims to achieve comprehensive enhancements in both machinability and sustainability. The optimization objectives encompass improving surface quality, dimensional accuracy, material removal rate, and specific energy consumption during the micro-milling process for thin-walled micro-structures. Among them, NSGA-III achieves the best optimization results. Under conditions in which cutting energy consumption and processing efficiency are very close, the optimization outcomes based on NSGA-III lead to the best machining quality, including the minimum surface roughness and dimensional errors, and the largest surface fractal dimension. The optimal combination of micro-milling parameters is n = 28,800 rpm, f_z = 2.6 μm/t, and a_p = 62 μm.

Keywords: thin-walled micro parts; micro-milling; machinability; sustainability; multi-objective optimization

check for updates

Citation: Wang, P.; Bai, Q.; Cheng, K.; Zhao, L.; Zhang, Y. Multi-Objective Optimization of Micro-Milling Parameters—The Trade-Offs between Machining Quality, Efficiency, and Sustainability in the Fabrication of Thin-Walled Microstructures. *Appl. Sci.* **2023**, *13*, 9392. https://doi.org/10.3390/app13169392

Academic Editor: Guian Qian

Received: 28 July 2023
Revised: 13 August 2023
Accepted: 16 August 2023
Published: 18 August 2023

1. Introduction

The optimization of process parameters is regarded as a crucial means to improve machining performance, the results of which are dependent on the optimization objectives and methods. In prior research, optimization objectives have mainly focused on technical requirements and production efficiency, while energy consumption and environmental degradation have often been neglected [1]. Sustainable manufacturing technology presents new challenges for researchers in the context of carbon peaking and carbon neutrality.

As some of the main energy-consuming pieces of equipment in the manufacturing industry, the primary form of energy consumption for machine tools is electricity. There are approximately 8 million units of machine tools in China, indicating significant potential for energy saving. Therefore, balancing machining quality and efficiency while reducing the energy consumption of machine tools has gradually become a prominent research focus in the manufacturing industry [2]. Hu et al. employed the simulated annealing algorithm to optimize the spindle speed and feed rate in the single-point turning process, focusing on machining energy consumption while considering cutting force, cutting power, and surface roughness as constraints. The optimized process parameters resulted in a 19.28% reduction in energy consumption [3]. Feng et al. conducted simultaneous optimization of toolpaths and process parameters during the drilling process using NSGA-II, with energy consumption, machining time, and surface roughness as the optimization objectives. The experimental results demonstrated that this approach reduced the energy consumption and machining time by 57.7% and 66.4%, respectively, while significantly improving the surface quality [4]. In order to achieve a balance between milling sustainability, production efficiency, and machining quality, Yan et al. employed the response surface methodology in combination with the grey relational analysis method to optimize the spindle speed, feed rate, cutting depth, and cutting width in the milling process. The optimized process parameters were found to enhance both the material removal rate and surface quality while reducing energy consumption by 18.1% [5]. Zhang et al. conducted an optimization study on micro-milling process parameters, including spindle speed, feed per tooth, and cutting depth, with the objective of minimizing power consumption, extending tool life, improving surface roughness, and enhancing tool holder strength. The cuckoo search algorithm and grey wolf optimization algorithm were used to enhance global search capabilities. As a result, a 7.89% reduction in energy consumption was achieved [6]. Wang et al. conducted optimizations of the energy consumption and machining time in milling processes. The results indicated that the optimization outcomes based on the improved ant colony algorithm surpassed those obtained with NSGA-II, yielding higher productivity at the same energy consumption level [7]. Rational parameter selection can lead to comprehensive improvements in machinability and sustainability at lower costs. The current research has mostly focused on conventional machine tools, with limited studies on micro-milling.

The established energy consumption model is the basis for analyzing energy characteristics, improving energy efficiency, and minimizing energy consumption in machining. Currently, CNC machine tool energy consumption models can be generally categorized into three types: material-removal-rate-based models, specific process-parameter-based models, and cutting-force-based models [8]. The power consumption model based on material removal rates is simple in form and has clear physical significance. However, research has shown that different combinations of process parameters can yield different energy consumption values at the same material removal rate [9–12]. The variation in process parameters is a significant factor influencing the energy consumption of machine tools [13]. In addition to regression methods, artificial intelligence is also employed to establish the relationship between process parameters and machine tool power consumption [14]. However, these models lack physical interpretability, and the energy consumption model based on artificial intelligence algorithms relies on extensive testing data [15,16]. Due to factors such as tool wear, tool deflection, and workpiece material properties, accurately modeling cutting forces presents challenges [17]. Consequently, establishing a cutting power model based solely on cutting forces may deviate from actual results [18,19].

Micro-milling is a key technology for manufacturing precision components and finds widespread applications in various fields. Due to the influence of size effects, micro-milling has a higher specific energy consumption, which is particularly pronounced when processing difficult-to-machine materials such as Ti-6Al-4V [20]. Although many researchers have analyzed the power consumption characteristics of cutting processes and established corresponding power consumption models using various methods, most of these models are focused on general-machining machine tools and still have certain limitations, particularly a lack of in-depth explanations of the power consumption mechanisms and unclear physical meanings [21–24]. Therefore, it is crucial to determine a more generalized process-parameter-based power consumption modeling approach for micro-milling processes.

To reduce the energy consumption in micro-milling processes, this paper carried out power consumption modeling using power flow analysis. Based on the established model, the optimization of process parameters was undertaken with the objective of reducing the specific energy consumption, considering constraints on machining quality and efficiency, thereby achieving comprehensive improvements in machinability and sustainability.

2. Materials and Methods

2.1. Power Consumption Modeling

During the machining process, all the various forms of energy generated by machine tools are derived from electrical energy, including mechanical energy, sound energy, and thermal energy. Therefore, in this paper, when measuring the power consumption of various energy-consuming subsystems of machine tools, only the electrical energy consumption was measured.

In this paper, the machine tool power consumption is split into several components, and their sum is the total power consumption of the machining process, which can be illustrated with Equation (1).

$$P_{\text{total}} = \sum_{i=1}^{I} P_{\text{fix}-\text{i}} + \sum_{j=1}^{J} P_{\text{var}-\text{j}}(n, f_z, a_p) \tag{1}$$

where P_{total} represents the total power consumption of the machine tool; $P_{\text{fix}-\text{i}}$ represents the power consumption of the i-th fixed energy consumption subsystem; $P_{\text{var}-\text{j}}$ represents the power consumption of the j-th variable energy consumption subsystem; I represents the number of fixed energy consumption subsystems; J represents the number of variable energy consumption subsystems; n represents the spindle speed; f_z represents the feed per tooth; and a_p represents the cutting depth.

The power consumed by each fixed energy consumption subsystem can be considered as a constant value that does not vary with machining states. The power consumed by each variable energy consumption subsystem is influenced by process parameters, cutting tools, and workpiece materials. Therefore, this paper focused on analyzing and establishing power consumption models for variable energy consumption subsystems. The Kern EVO micro-milling machining center was the research object, as shown in Figure 1.

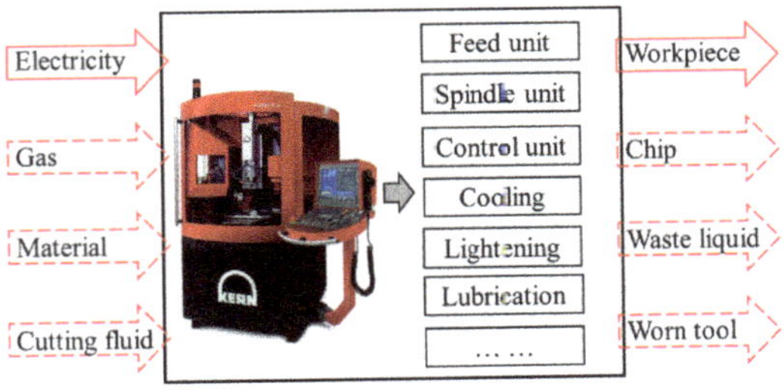

Figure 1. Energy consumption boundary of Kern Evo micro-milling machine tool.

On this basis, the total power consumed by the machine tool can be further decomposed as Equation (2).

$$P_{\text{total}} = P_{\text{i}} + P_{\text{m}} = P_{\text{b}} + P_{\text{a}} + P_{\text{s}} + P_{\text{f}} + P_{\text{m}} \tag{2}$$

where P_{b} represents the basic power of the control systems, lighting, and air pumps; P_{i} represents idle power consumption; P_{a} represents the auxiliary power of the cooling, lubrication, chip collection, and tool-changing systems; P_{s} represents the non-machining spindle power; P_{f} represents the non-machining feed power; and P_{m} represents the material removal power.

The machine tool power varies with different machining process stages, as shown in Figure 2. It can be observed that the tool change process has a short duration and low power consumption, which can be neglected if the tool change frequency is not high. Upon activating the cutting lubrication system, there is a significant increase in the machine tool power. Furthermore, it can be seen that different process stages exhibit distinct variations in cutting power, indicating that changes in process parameters lead to different power consumptions.

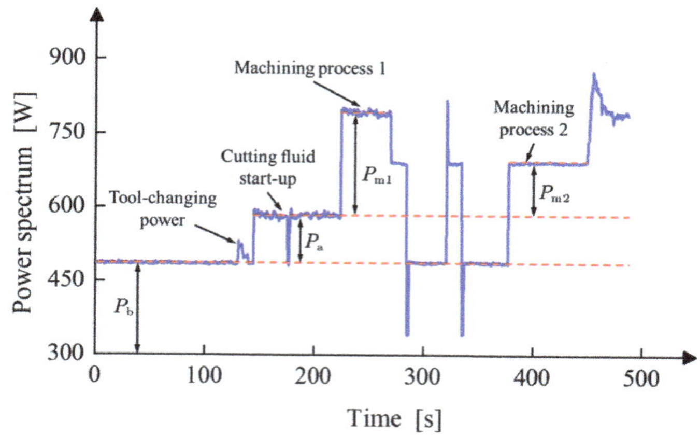

Figure 2. Power spectrum of different machining states.

2.1.1. Power Consumption Modeling in Feed and Spindle Subsystems

A schematic diagram of the feed unit motion structure for micro-milling machining equipment is illustrated in Figure 3. The power source for the feed system is synchronous permanent magnet motors, which transmit driving force (torque) to the worktable through couplings, bearings, and ball screws. For the horizontal feed drive unit, the load refers to the friction of the ball screw pair, and for the vertical feed drive unit, the effect of gravity should also be considered. It should be noted that the cutting force load is not included. Similarly, the load of the spindle system does not consider the influence of the cutting force torque.Synchronous permanent magnet motors convert input electrical energy into mechanical energy, and the motor's output power can be obtained via force analysis. The relationship between the motor's output torque and the forces on the ball screw, bearings, and couplings can be described using Equations (3)–(5).

$$T_{\mathrm{m}} = J_{\mathrm{m}} \frac{d\omega_{\mathrm{m}}}{dt} + C_{\mathrm{m}}\omega_{\mathrm{m}} + T_{\mathrm{c}} \tag{3}$$

$$T_{\mathrm{c}} = J_{\mathrm{c}} \frac{d\omega_{\mathrm{c}}}{dt} + C_{\mathrm{c}}\omega_{\mathrm{c}} + T_{\mathrm{bs}} + 2T_{\mathrm{b}} \tag{4}$$

$$T_{\mathrm{bs}} = J_{\mathrm{bs}} \frac{d\omega_{\mathrm{bs}}}{dt} + C_{\mathrm{bs}}\omega_{\mathrm{bs}} + T_{\mathrm{t}} \tag{5}$$

where T, J, C, and ω represent the output torque, moment of inertia, viscous damping, and angular velocity, and subscripts m, c, bs, b, and t represent the motor, coupling, ball screw, bearing, and worktable.Under the assumption of a negligible angular velocity difference, the relationship between the motor output torque and the workbench output torque can be expressed as Equation (6).

$$T_{\mathrm{m}} = J \frac{d\omega}{dt} + C \cdot \omega + T \tag{6}$$

where T, J, and C represent the output torque, moment of inertia, viscous damping, and angular velocity of the feed system, which can be expressed as Equation (7).

$$\begin{cases} J = J_m + J_c + J_{bs} \\ C = C_m + C_c + C_{bs} \\ T = T_t + 2T_b \end{cases} \tag{7}$$

After obtaining the motor's output torque, the power output of the driving motor can be represented as Equation (8).

$$P_{output} = T_m \cdot \omega = J \cdot \frac{d\omega}{dt} \cdot \omega + C \cdot \omega^2 + T \cdot \omega \tag{8}$$

Apart from the motor output power, permanent magnet synchronous motors also have internal energy losses, which can generally be classified into the following categories: copper loss, iron loss, mechanical loss, additional loss, and electromagnetic loss, as shown in Figure 4.Prior research has indicated that the power loss of a motor can be represented as a quadratic function of the motor's rotational speed, as shown in Equation (9) [25].

$$P_{loss} = k\omega^2 + b\omega + P_c \tag{9}$$

where k and b are coefficients determined by the motor itself, and P_c represents the fixed power loss.Therefore, the expression for the power consumption of the feed system can be obtained by combining Equations (8) and (9).

$$P_{feed} = P_{output} + P_{loss} = F_0 \cdot \alpha \cdot \omega + F_1 \cdot \omega^2 + F_2 \cdot \omega + F_3 \tag{10}$$

where α represents the angular acceleration, and the coefficients can be expressed as Equation (11).

$$\begin{cases} F_0 = J \\ F_1 = C + k \\ F_2 = T + b \\ F_3 = P_c \end{cases} \tag{11}$$

In the stable machining process, the feed speed is constant (i.e., the angular acceleration is 0); thus, the power consumption can be expressed as:

$$P_{feed-s} = F_1 \cdot \omega^2 + F_2 \cdot \omega + F_3 \tag{12}$$

Through a similar force analysis process as that for the feed unit, the power consumption of the electric spindle can be expressed as follows:

$$P_{spindle-s} = S_1 \cdot n^2 + S_2 \cdot n + S_3 \tag{13}$$

where S_1 represents the viscous damping of the electric main spindle; S_2 represents the output torque of the electric main spindle; and S_3 represents the power loss of the electric main spindle.

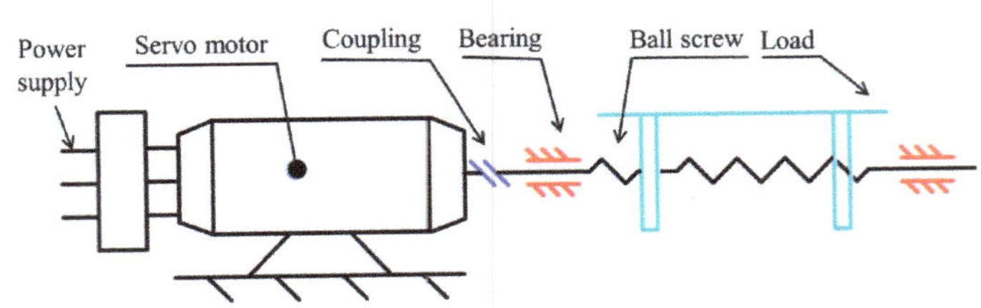

Figure 3. Structural diagram of feed system.

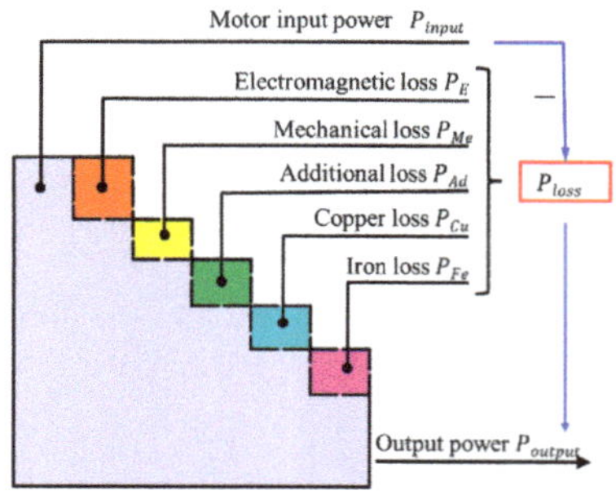

Figure 4. The power loss of drive motor.

2.1.2. Material Removal Power Modeling

Material removal power refers to the power consumed in the interaction between cutting tools and workpiece materials to achieve specific shapes and meet certain technical requirements. There are several factors that influence material removal power, making accurate modeling challenging. The power consumption in the cutting process is directly influenced by process parameters. Therefore, establishing the material removal power based on specific process parameters offers the advantages of simplicity, directness, and accuracy, facilitating the further optimization of process parameters. In this paper, both exponential and polynomial models are used to fit and analyze cutting power, as shown in Equations (14) and (15), respectively, and a comparison of their merits is presented.

$$P_{\mathrm{m}} = a \cdot n^{c_1} \cdot f_z^{c_2} \cdot a_p^{c_3} \tag{14}$$

$$\begin{aligned} P_{\mathrm{m}} = p_0 + p_1 \cdot n + p_2 \cdot a_p + p_3 \cdot f_z + \\ p_4 \cdot n \cdot a_p + p_5 \cdot n \cdot f_z + p_6 \cdot a_p \cdot f_z \end{aligned} \tag{15}$$

where n, f_z, and a_p represent the spindle speed, feed per tooth, and depth of cut, respectively, and a, c_1, c_2, c_3, p_0, p_1, p_2, p_3, p_4, p_5, and p_6 are coefficients that are closely related to the cutting tools, materials, and process conditions.

2.2. Description of the Multi-Objective Optimization Problem

In the material removal process, improving a specific technical index by adjusting the process parameters may lead to the deterioration of performance in other areas. Therefore, the objective of this study was to balance the relationship between sustainability, machining quality, and efficiency in micro-milling processes, which is illustrated in Figure 5.

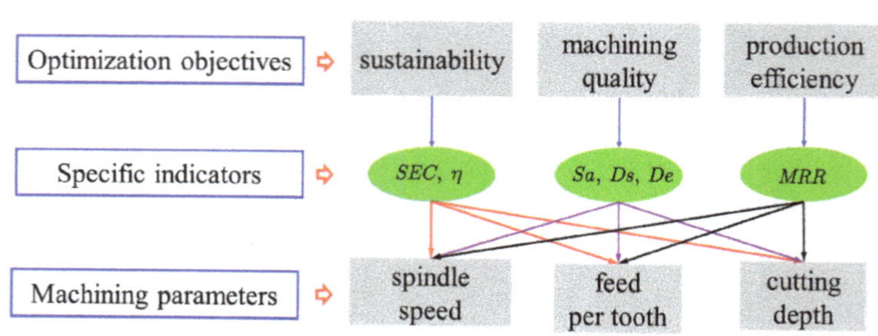

Figure 5. The multi-objective optimization problem description.

In this study, the optimization objectives included surface roughness (S_a), surface fractal dimension (D_s), and size error (D_e) as measures of machining quality; the material removal rate (MRR) considered for evaluating machining efficiency; and the specific cutting energy consumption (SCE) employed for energy efficiency assessment. Their definitions are shown in Equations (16)–(20).

$$S_a = \iint_S \Phi(x,y) - \xi(x,y)|dxdy/S \tag{16}$$

where $\Phi(x,y)$ denotes the height information of the surface, and $\xi(x,y)$ corresponds to the reference plane of the surface.

$$D_s = \frac{\ln N(\varepsilon)}{\ln(\varepsilon^{-1})} \tag{17}$$

where ε represents the side length of the box, and $N(\varepsilon)$ represents the number of boxes. The value of the surface fractal dimension represents the richness of details on the surface, reflecting the complexity and finesse of the surface contour in space, as well as its ability to fill space. In comparison with parameters of surface roughness, the surface fractal dimension is more sensitive to the absence of surface microstructures. Therefore, selecting the surface fractal dimension to characterize surface quality holds a certain significance [26,27].

$$D_e = |D_a - D_i| \tag{18}$$

where D_e is the actual dimension, D_a is the actual dimension, and D_i is the design dimension. The dimensional errors (D_e) were measured using pixel analysis of the SEM-measured picture. In comparison with image scales, three positions on the thin-walled microstructure were selected for averaging; thus, the actual dimensions were determined, followed by subtracting the design dimensions to obtain the dimensional errors.

$$MRR = a_p \cdot a_e \cdot f \tag{19}$$

where a_p, a_e, and f are the cutting depth, cutting width, and feed speed, respectively.

$$SCE = \frac{E_{\text{total}}}{V_{\text{material}}} = \frac{P \cdot T}{MRR \cdot T} = \frac{P}{MRR} \tag{20}$$

where SCE, E_{total}, V_{material}, T, and P are the specific cutting energy, cutting energy, volume of removal material, machining time, and machining power, respectively.

The three process parameters to be optimized are the spindle speed n, feed per tooth f_z, and cutting depth a_p. Considering the main technical requirements of thin-walled microstructures, the optimal combination of process parameters can be obtained with the constraints shown in Equations (21) and (22).

$$\begin{cases} n_{\min} \le n \le n_{\max} \\ f_{z\min} \le f_z \le f_{z\max} \\ a_{p\min} \le a_p \le a_{p\max} \end{cases} \tag{21}$$

$$\begin{cases} \min Sa(n, f_z, a_p) \\ \max Ds(n, f_z, a_p) \\ \min De(n, f_z, a_p) \\ \max MRR(n, f_z, a_p) \\ \min SEC(n, f_z, a_p) \end{cases} \tag{22}$$

To eliminate the negative impacts caused by the improper selection of process parameters, it is essential to adopt appropriate optimization methods, which generally fall into two categories: traditional methods and intelligent methods. Traditional methods typically rely on experimental design and gradient-based search, the limitation of which lies in the fact that the convergence can be greatly influenced by the nonlinearity of the objective function and initial value settings and the global optimality is hard to ensure. Therefore, in recent years, intelligent optimization algorithms have been widely applied to parameter optimization in machining, with evolutionary algorithms being the most extensively used. The high dimensionality of the optimization objective space poses challenges in the optimization process. Therefore, in this paper, various process parameter optimization methods were utilized, and their results were compared, including NSGA-II (non-dominated genetic sorting algorithm-II), MOPSO (multi-objective particle swarm optimization), and NSGA-III (non-dominated genetic sorting algorithm-III) [28,29].

2.3. Experimental Design

The micro-milling experiments were conducted in the Kern Evo five-axis vertical machining center. The cutting tools had a diameter of 800 μm and a helix angle of 60 degrees. The workpiece material was a Ti-6Al-4V alloy, the main dimensions of which are shown in Figure 6. In order to analyze and validate the power consumption model proposed above, the Fluke 1735 three-phase power logger was employed to record the power characteristics during the machining process, as shown in Figure 7.

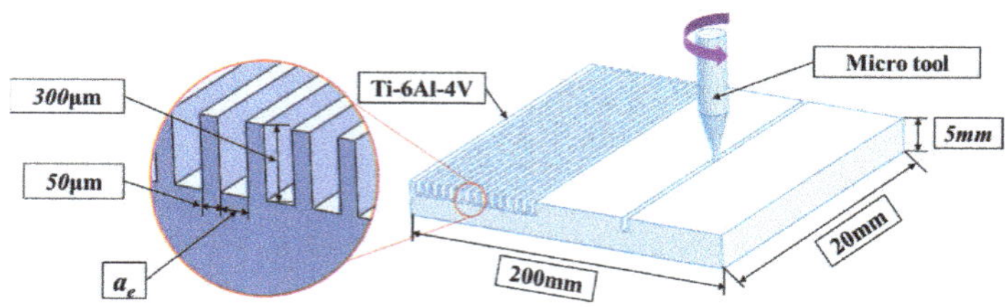

Figure 6. Main dimensions of workpiece.

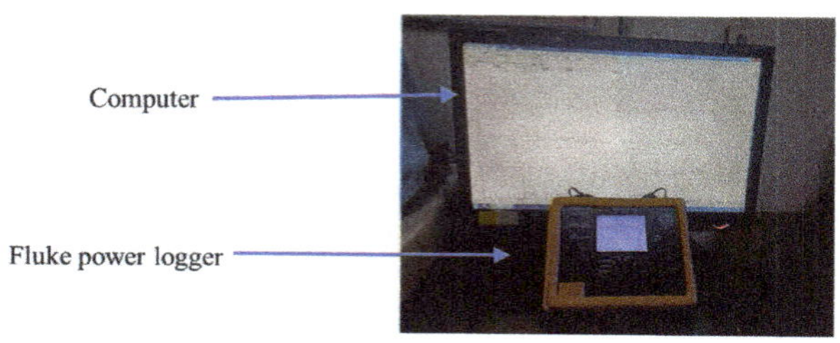

Figure 7. Measurement of power consumption.

Due to the feed and spindle power consumption only being influenced by single factors in the non-machining stage, an experiment was designed to validate the proposed power consumption models, and the corresponding parameters are shown in Table 1.

Table 1. Process parameters for non-machining power measurement.

f (mm/min)	n (rpm)	a_p (μm)
20, 120, 220, 320, and 420	0	0
0	10,000, 15,000, 20,000, 25,000, 30,000, 35,000, 40,000, and 45,000	0

The material removal power is influenced by the combination of process parameters. In order to analyze the effects of the process parameters on the machining quality, production efficiency, and energy efficiency, an orthogonal experiment was designed, and the corresponding levels are shown in Table 2. Ignoring the interactions between process parameters, the experimental design was conducted based on the L16 Taguchi experimental table [30], as shown in Table 3.

Table 2. Machining parameters used in orthogonal experiments.

Symbol	Process Parameters	Units	Level			
			1	2	3	4
A	Spindle speed (n)	r/min	15,000	25,000	35,000	45,000
B	Axial depth of cut (a_p)	μm	50	60	100	150
C	Feed per tooth (f_z)	μm/tooth	0.5	1.5	2.5	3.5

Table 3. Machining parameters and machining responses of the orthogonal experiments.

No.	Symbol	S_a (nm)	D_e (μm)	D_s	SCE (kJ/mm^3)	MRR (mm^3/min)	Level		
							A	B	C
1	A1B1C1	45	1.3	2.5110	15.94	0.60	15,000	50	0.5
2	A1B2C2	56	2.8	2.4620	4.50	2.16	15,000	60	1.5
3	A1B3C3	60	2.9	2.4468	1.67	6.00	15,000	100	2.5
4	A1B4C4	143	4.1	2.4108	0.82	12.60	15,000	150	3.5
5	A2B1C2	48	3.2	2.4960	15.50	1.20	20,000	60	0.5
6	A2B2C1	54	2.2	2.4730	6.15	3.00	20,000	50	1.5
7	A2B3C4	185	4.3	2.3663	1.28	15.00	20,000	150	2.5
8	A2B4C3	121	4.4	2.3903	1.38	14.00	20,000	100	3.5
9	A3B1C3	57	2.8	2.4565	10.27	2.80	25,000	100	0.5
10	A3B2C4	133	4.6	2.4478	2.31	12.60	25,000	150	1.5
11	A3B3C1	58	2.7	2.4500	4.15	7.00	25,000	50	2.5
12	A3B4C2	99	3.5	2.4346	2.49	11.76	25,000	60	3.5
13	A4B1C4	60	2.3	2.4441	8.22	5.40	30,000	150	0.5
14	A4B2C3	85	4.7	2.4346	4.09	10.80	30,000	100	1.5
15	A4B3C2	90	3.8	2.4291	4.10	10.80	30,000	60	2.5
16	A4B4C1	66	5.7	2.4391	3.52	12.60	30,000	50	3.5

3. Results and Discussions

For the non-machining experiments, the power results corresponding to different feed rates are shown in Figure 8. It can be observed that the power values exhibit an increasing trend with the increasing feed rate.

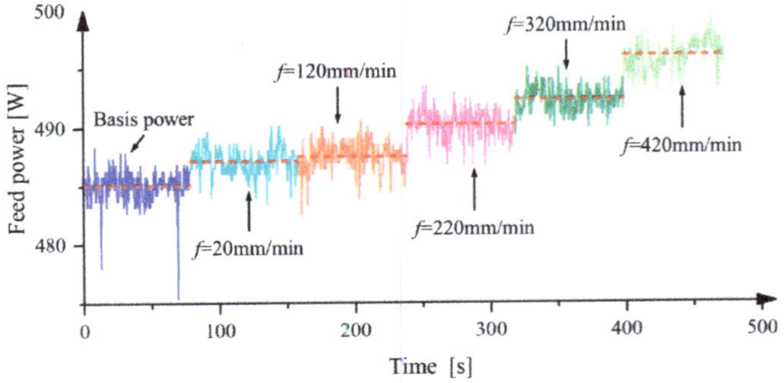

Figure 8. Non-machining power at different feed speeds.

To determine the model coefficients, a quadratic fitting was performed, and the trend of the power variation with the feed rate was obtained, as depicted in Figure 9.

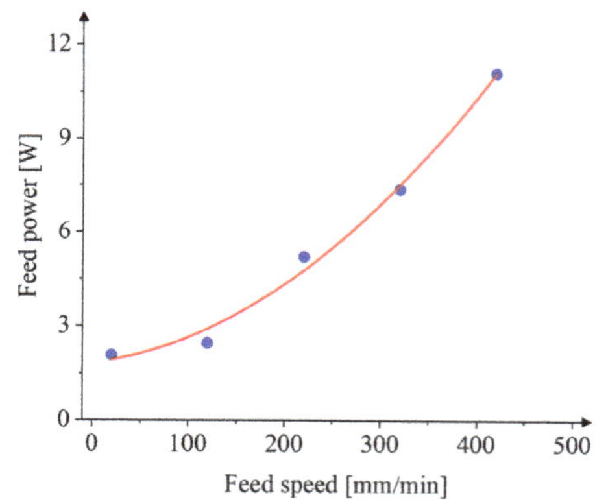

Figure 9. Variation in non-machining power with feed speed.

The result of the quadratic fitting is shown in Equation (23).

$$P_{\text{feed}-s} = 4.4 \times 10^{-5} \cdot f^2 + 0.0038 \cdot f + 1.83 \tag{23}$$

The variance (R^2), adjusted variance (R^2-adj), and prediction variance (R^2-pre) of the proposed model all exceeded 90%, as shown in Table 4, which indicates that the modeling method presented in this study demonstrates high prediction accuracy.

Table 4. Non-machining power model prediction accuracy.

Feed Unit			Spindle Unit		
R^2	R^2-pre	R^2-adj	R^2	R^2-pre	R^2-adj
99.21%	94.42%	98.43%	99.92%	99.82%	98.89%

The variation trend of the spindle power at different speeds is shown in Figure 10. It can be observed that, compared with the feed motion, the spindle rotational motion consumes higher power. Therefore, in the process of energy consumption optimization, special attention should be given to the spindle power.

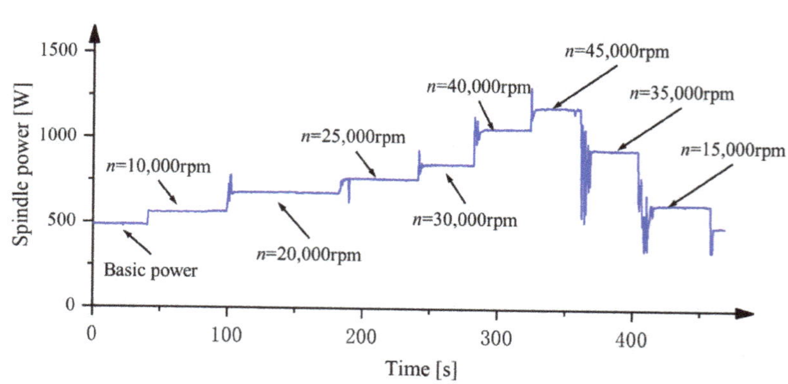

Figure 10. Non-machining power at different spindle speeds.

The power consumption of the spindle demonstrates a well-fitted quadratic relationship with the rotational speed, as depicted in Figure 11. The fitting result is presented in Equation (24). The variance (R^2), adjusted variance (R^2-adj), and prediction variance (R^2-pre) of the proposed fitting model all exceeded 98%, as shown in Table 4, which indicates that the proposed modeling method demonstrates good prediction accuracy.

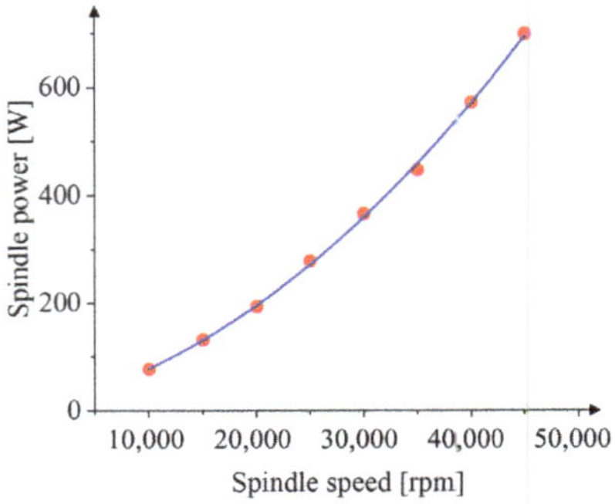

Figure 11. Variation in non-machining power with spindle speed.

$$P_{\text{spindle-s}} = 2.37 \times 10^{-7} \cdot n^2 + 0.00461 \cdot n + 1.3 \quad (24)$$

To facilitate the fitting of the relationships between the process parameters and material removal power, Equation (14) was further manipulated into Equation (25). A linear regression approach can be employed to obtain the coefficients for each term.

$$\ln P_{\text{mr}} = \ln a + c_1 \ln n + c_2 \ln f_z + c_3 \ln a_p \quad (25)$$

Under each set of process parameters, the total power P_{tatal} can be measured when performing each experiment; the machine base power P_b can be measured when the machine tool is powered on but without performing any movement or processing, as shown in Figure 2; the spindle power P_s and feed power P_f can be calculated using Equations (23) and (24), respectively; and after determining the power component values above, the material removal power can be calculated using Equation (26). The corresponding values are presented in Table 5.

$$P_m = P_{\text{total}} - (P_b + P_a + P_s + P_f) \quad (26)$$

To achieve the best predictive accuracy, a comparison between the exponential fitting and polynomial fitting models was performed. The results of the exponential and polynomial fittings are shown in Equation (27) and Equation (28), respectively.

$$P_{\text{mr}} = 14.92 \cdot n^{0.0613} \cdot f_z^{0.1214} \cdot a_p^{0.0708} \quad (27)$$

$$\begin{aligned} P_{\text{mr}} = {} & 10.18 + 6 \times 10^{-5} \cdot n + 0.153 \cdot ap + 1.98 \cdot fz + \\ & 1 \times 10^{-6} \cdot n \cdot a_p - 1.9 \times 10^{-5} \cdot n \cdot f_z - 0.0034 \cdot a_p \cdot f_z \end{aligned} \quad (28)$$

To compare the goodness of fit and predictive accuracy of the two fitting models, the variance (R^2), adjusted variance (R^2-adj), predictive variance (R^2-pre), Akaike information criterion (AIC), and Bayesian information criterion (BIC) values were obtained, as shown in Table 6.

Table 5. Values of different power components in orthogonal experiments.

No.	P_{tatal} (W) (Measured)	P_b (W) (Measured)	P_s (W) (Calculated)	P_f (W) (Calculated)	P_m (W) (Calculated)
1	645.4	486	127	1.9	33.5
2	648.1			2.1	36.0
3	653.3			2.4	40.9
4	659.3			2.7	46.6
5	793.9		271	2.4	36.5
6	795.7			2.0	34.7
7	802.5			3.8	44.7
8	806.9			3.0	46.9
9	965.2		440	3.8	35.4
10	969.0			5.4	39.6
11	970.7			2.1	42.6
12	976.2			2.7	45.5
13	1221.4		692	7.4	40.0
14	1202.3			4. 9	38.7
15	1224.0			3.1	42.9
16	1228.5			2.1	46.4

Table 6. The goodness of fit of P_m.

	R^2	R^2-adj	R^2-pre	AIC	BIC
Exponential model	0.8589	0.8236	0.7300	−34.32	−36.46
Polynomial model	0.9185	0.8641	0.6755	89.53	75.14

It is evident that selecting different criteria as the goodness-of-fit measures leads to different conclusions. In comparison with the exponential model, the polynomial fitting model exhibits higher values for R^2 and R^2-adj but a lower value for R^2-pre. However, the differences are not very significant. To better explain the selection rationale, the AIC and BIC values of both models were also compared. The AIC and BIC values of the exponential fitting model are both smaller than the polynomial fitting model, indicating that the exponential model is superior from both predictive and fitting perspectives.

After obtaining models for the spindle power, feed power, and material removal power, the variable power could be obtained using Equation (29).

$$P_{var} = P_s + P_f + P_m \tag{29}$$

From the range analysis, it can be observed that the order of the process parameters' influences on the variable power is as follows: spindle speed, feed per tooth, and cutting depth. The range value of the spindle speed is significantly larger than the range values of the feed per tooth and cutting depth, which indicates that during micro-milling processes, the cutting energy consumed by the spindle unit is the highest, as shown in Table 7. The main effects of the process parameters on the variable power also support this result, as shown in Figure 12.

Table 7. Range analysis of P_{var}.

	n (rpm)	f_z (μm/t)	a_p (μm)
R_j	572.9	8.7	6.2
Ranking	1	2	3

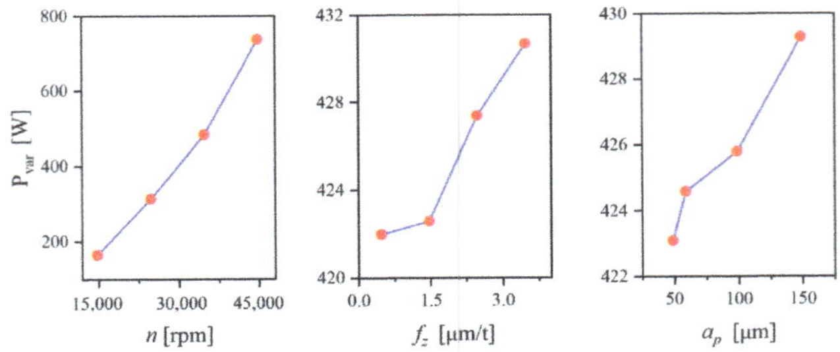

Figure 12. Main effects of process parameters on variable power.

The surface topographies are shown in Figure 13, which were measured with a white-light interferometer. The measurement range was 396 μm by 396 μm. The indicators characterizing the micro-milling quality, including S_a, D_s, and D_e, are presented in Table 3. The corresponding regression models are shown in Equation (30), and the respective goodness-of-fit indices are listed in Table 8.

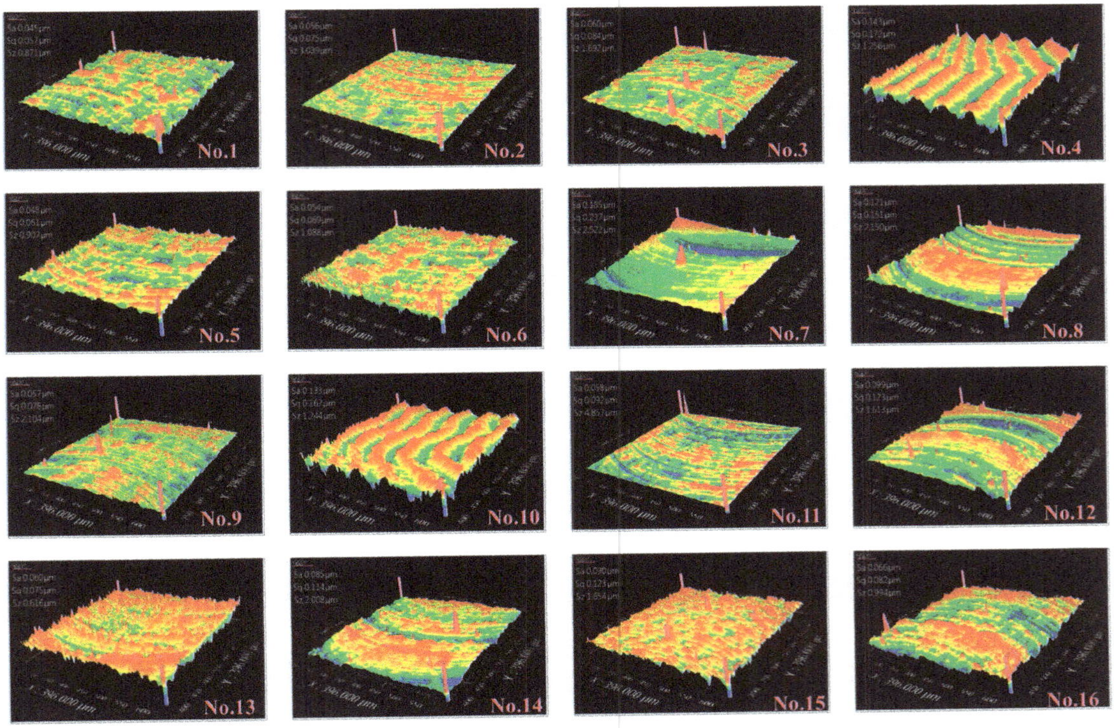

Figure 13. Surface topographies of the machined surface.

Table 8. Goodness-of-fit of machining quality indices.

S_a		D_s		D_e	
R^2	R^2-adj	R^2	R^2-adj	R^2	R^2-adj
77.71%	72.14%	81.27%	76.59%	67.47%	59.33%

$$\begin{cases} S_a = 1.8221 \times n^{0.0830} \times f_z^{0.3404} \times a_p^{0.6230} \\ D_s = 2.1839 \times n^{0.0059} \times f_z^{0.0125} \times a_p^{0.0181} \\ D_e = 0.0247 \times n^{0.3490} \times f_z^{0.3076} \times a_p^{0.2630} \end{cases} \quad (30)$$

The main effects of the process parameters on the surface roughness, fractal dimension, and dimension error are shown in Figures 14–16. It can be observed that the surface roughness and fractal dimension exhibit distinct opposite trends because the fractal dimension characterizes the richness of details on the surface. The larger the fractal dimension, the more complex and intricate the structures are. Conversely, the smaller the fractal dimension, the fewer irregular structures there are, and the rougher the structure becomes.

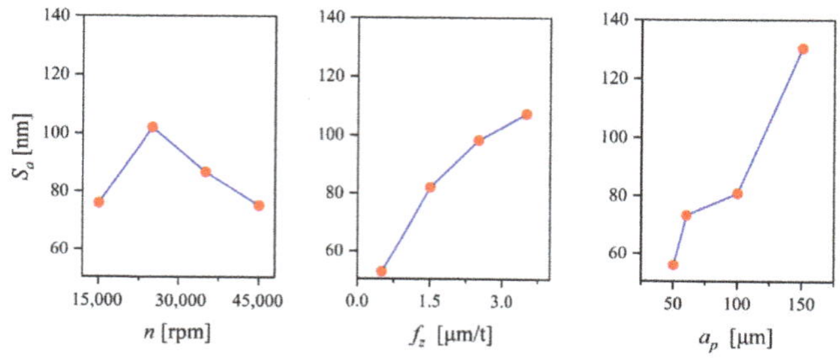

Figure 14. Main effects of process parameters on S_a.

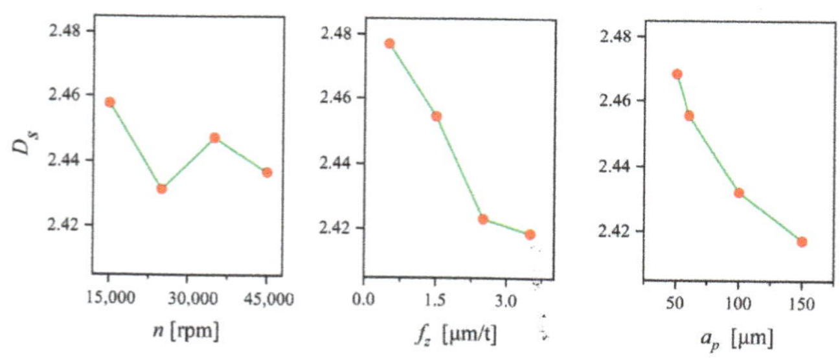

Figure 15. Main effects of process parameters on D_s.

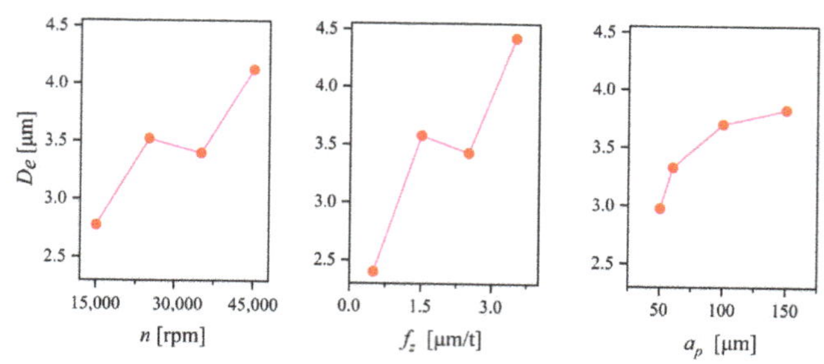

Figure 16. Main effects of process parameters on D_e.

The dimensional error shows an overall increasing trend with the enlargement of the process parameters. This is due to the fact that with the increase in the feed per tooth and cutting depth, the cutting forces exhibit an increasing trend. This leads to greater deformation of thin-walled micro-structures in the vertical feed direction, ultimately resulting in an increase in dimensional error. Conversely, as the spindle speed increases, the heat generated during cutting also increases, making the workpiece material prone to thermal deformation. Consequently, this leads to an augmentation in dimensional error. The reasons for small deformation errors are as follows: The micro-milling force is relatively small, and previous studies have shown that the average cutting force for the micro-milling of titanium alloys is within 5 N [31]; the positioning accuracy of the KERN EVO machining center is 1 μm, with a repeatability accuracy of $\pm$ 0.5 μm, exhibiting excellent machining precision; and Ti-6Al-4V possesses high strength, and under the influence of relatively small cutting forces, noticeable deformation errors are not likely to occur.

The main effects of the process parameters on the SCE are shown in Figure 17. It can be observed that the SCE significantly decreases with the increase in feed per tooth and cutting depth because their impacts on the material removal rate do not significantly affect power consumption. However, the multiplication of the spindle speed leads to a rapid increase in power consumption.

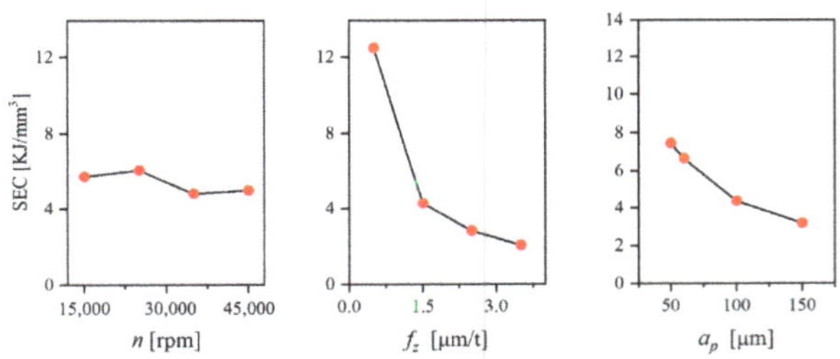

Figure 17. Main effects of process parameters on SCE.

The relationship between the SCE and MRR is illustrated in Figure 18, with a Pearson correlation coefficient of −0.793, which exceeds 0.7, indicating a significant negative correlation. This implies that enhancing the machining efficiency can reduce specific energy consumption.

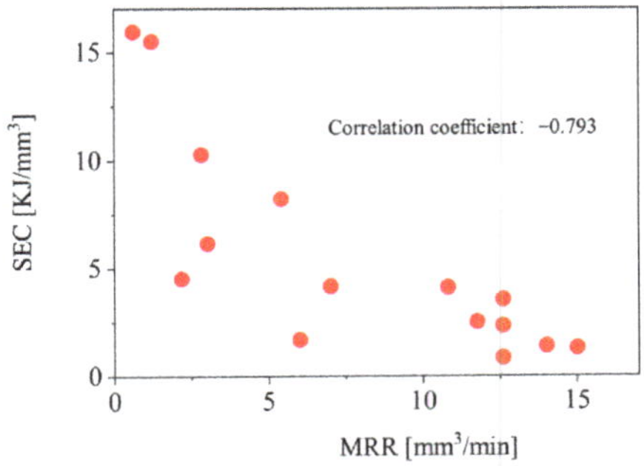

Figure 18. Correlation between MRR and SCE.

In order to overcome the problem of intelligent optimization algorithms easily getting trapped in local optima, this paper investigated and compared the optimal combination of process parameters using three evolutionary algorithms. Since NSGA-II and MOPSO perform poorly when dealing with four or more optimization objectives, the optimization objectives were first compressed. To retain and interpret the original variable information as much as possible, principal component analysis (PCA) was applied to the three indicators representing the machining quality. First, the standardization of S_a, D_s, and D_e was performed according to Equation (31).

$$\begin{aligned} X'_{mn} &= \left(X_{mn} - \overline{X}_m\right) / \sigma_m \\ \overline{X}_m &= \sum_{n=1}^{p} X_{mn} / p \\ \sigma_m &= \sqrt{\sum_{n=1}^{p} \left(X_{mn} - \overline{X}_m\right)^2 / p} \end{aligned} \tag{31}$$

where X'_{mn} represents the n-th sampled value of the m-th objective function after normalization; X_{mn} represents the n-th sampled value of the m-th objective function; p represents the total number of samples; and $\overline{X}_m$ and σ_m represent the mean and standard deviation, respectively.

The surface roughness, fractal dimension, and dimensional error correlation coefficients are shown in Table 9. It can be observed that the surface quality indices, S_a and D_s, exhibit a strong negative correlation, while the correlation between surface quality and dimensional error is relatively weak.

Table 9. Correlations between the optimization objectives.

	S_a	D_s	D_e
S_a	1	−0.832	0.579
D_s	−0.832	1	−0.628
D_e	0.579	−0.628	1

Bartlett's sphericity test and the Kaiser–Meyer–Olkin (KMO) sampling adequacy test were conducted to assess the suitability of PCA, as shown in Table 10. The results indicate that Bartlett's sphericity test was significant (p = 0.000 < 0.05), and the KMO value was 0.681, which falls between 0.6 and 0.7. Therefore, PCA was appropriate.

Table 10. KMO and Bartlett's tests.

Sampling Adequacy KMO Measure		**0.681**
Bartlett's test of sphericity	Approximate chi-square	22.351
	df	3
	Sig.	0.000

Table 11 presents the common factor variance matrix, revealing that the S_a and D_s indicators for surface quality exhibit sufficient information extraction, and the information extraction content of D_e is relatively small.

Table 11. Common factor variance.

	Initial Variance	**Extracted Variance**
S_a	1.000	0.834
D_s	1.000	0.867
D_e	1.000	0.665

Higher eigenvalues indicate greater potential energy and are more crucial for the analysis. In Figure 19, only the first eigenvalue is greater than one, which indicates only one factor is needed to extract from the original features.

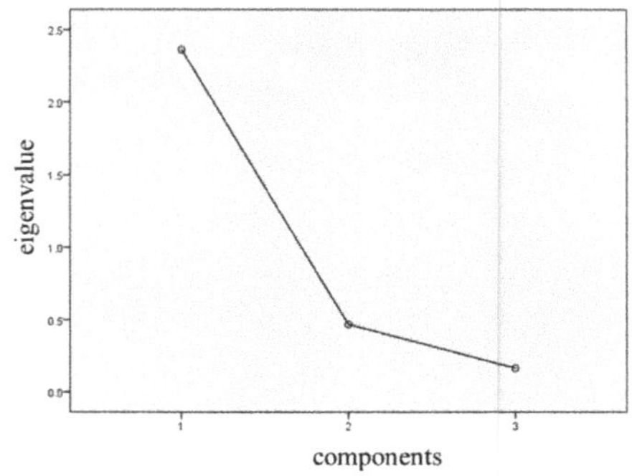

Figure 19. Scree plot of PCA results.

The explained total variance is shown in Table 12. The variance contribution of the first component is 78.851%, allowing for the extraction of original features with minimal information loss.

Table 12. Explained total variance.

No.	Initial Eigenvalue			Extract Loading Sum of Squares		
	Eigenvalue	Initial Variance %	Cumulative Variance %	Eigenvalue	Initial Variance %	Cumulative Variance %
1	2.366	78.851	78.851	2.366	78.851	78.851
2	0.470	15.654	94.505			
3	0.165	5.495	100.000			

The factor loading matrix is shown in Table 13, which illustrates the contributions of the principal components to each variable. According to the principal component matrix, the linear combinations of machining quality variables can be obtained as Equation (32). This principal component can be interpreted as the machining quality.

Table 13. Factor loading matrix.

S_a	D_s	D_e
0.386	−0.394	0.345

$$P_1 = 0.386(Sa) - 0.394(Ds) + 0.345(De) \tag{32}$$

Therefore, the objective function optimized with NSGA-II and MOPSO can be expressed as Equation (33), and the objective function optimized with NSGA-III can be expressed as Equation (34).

$$\min f_{1i}(X) = \begin{cases} f_1(X) = P_1 \\ f_2(X) = MRR^{-1} \\ f_3(X) = SEC \end{cases} \tag{33}$$

$$\min f_{2i}(X) = \begin{cases} f_1(X) = Sa \\ f_2(X) = Ds^{-1} \\ f_3(X) = De \\ f_4(X) = MRR^{-1} \\ f_5(X) = SEC \end{cases} \tag{34}$$

The decision variables can be represented as X = $[n, f_z, a_p]^T$. The ranges of the process parameters are as follows: 50 μm $\leq a_p \leq$ 150 μm, 15,000 rpm $\leq n \leq$ 45,000 rpm, and 0.5 mm/t $\leq f_z \leq$ 3.5 mm/t. The initial population size and number of iterations for each optimization algorithm were set as 500 and 1500, respectively. A membership function was used to select results from the Pareto solution set, as shown in Equation (35).

$$p_i = \frac{f_{i\max} - f_i}{f_{i\max} - f_{i\min}} \tag{35}$$

where $f_{i\max}$ and $f_{i\min}$ represent the maximum and minimum values of the i-th objective function, respectively, and f_i denotes the value of the i-th objective function.

The optimal solution is selected by maximizing the average p-value, as shown in Equation (36).

$$P = \frac{1}{N}\sum_{i=1}^{N} p_i \tag{36}$$

The optimization results of process parameters with different methods are listed in Table 14, and the corresponding experimental validations were performed, which can be observed in Figures 20 and 21, respectively. The results obtained using NSGA-III demonstrate a better trade-off between machinability and sustainability compared with the results using NSGA-II and MOPSO. It can be observed that the specific energy consumption and processing efficiency of the three optimization algorithms are very close. However, optimization results based on NSGA-III can achieve better machining quality, including lower surface roughness, larger surface fractal dimensions, and smaller dimensional errors.

Table 14. Optimization results.

	n	fz	a_p	S_a	D_s	D_e	MRR	SCE
NSGA-II	17,760	3.3	98	65	2.4860	3.5	9.37	0.960
MOPSO	21,000	3.2	88	59	2.4980	3.6	9.46	1.110
NSGA-III	28,800	2.6	62	50	2.5010	2.0	9.36	1.126

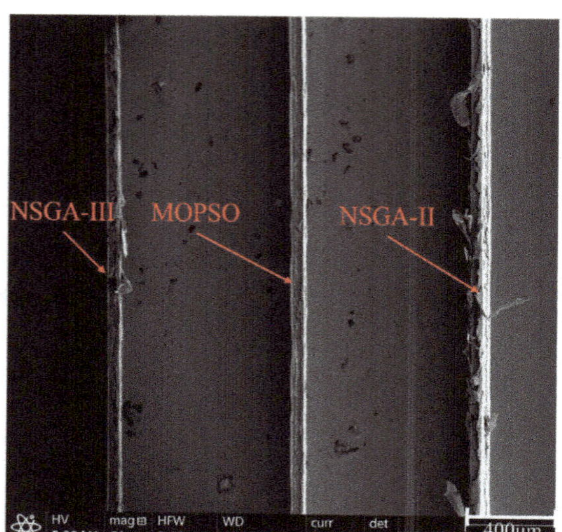

Figure 20. Dimensions of the thin-walled micro parts.

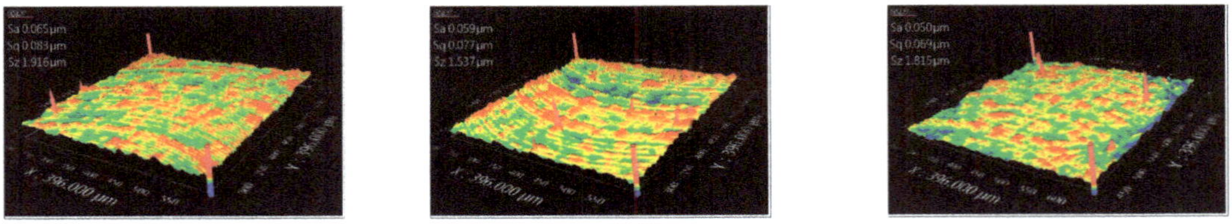

(**a**) (**b**) (**c**)

Figure 21. Surface topographies of the thin-walled micro parts: (**a**) NSGA-II; (**b**) MOPSO; and (**c**) NSGA-III.

4. Conclusions

An optimization strategy for micro-milling process parameters based on multi-objective evolutionary algorithms was proposed. The optimization objectives including machining quality, machining efficiency, and specific cutting energy were trade-offs. The obtained process parameters were validated using micro-milling experiments. The specific conclusions are as follows:

(1) The power consumption characteristics of the micro-milling process were analyzed, and a process-parameter-based power consumption model was established. The analysis revealed that the power consumption of non-machining feed motion exhibits a quadratic relationship with the feed rate, while the non-machining spindle power consumption shows a quadratic relationship with the spindle speed. The spindle speed exerts the greatest influence on the variable power consumption.

(2) The coefficient between SCE and MRR was −0.793, which indicates that enhancing machining efficiency can simultaneously improve machining sustainability.

(3) Among the three indicators characterizing machining quality, surface roughness and surface fractal dimension exhibited a significant negative correlation (with a correlation coefficient of −0.832). However, their correlations with dimensional error were relatively small, being 0.57 and −0.628, respectively. After performing PCA, the variance extraction percentages for S_a, D_s, and D_e all exceeded 50%, indicating a good level of information condensation.

(4) The optimization results of the process parameters demonstrate that the NSGA-III-based optimization method yields a better trade-off between machinability and sustainability, indicating that NSGA-III exhibits superior performance in handling multi-objective optimization problems and can obtain globally optimal parameters. The optimal combination of micro-milling parameters was n = 28,800 rpm, f_z = 2.6 μm/t, and a_p = 62 μm.

Author Contributions: Conceptualization, P.W. and Q.B.; methodology, P.W.; software, P.W.; validation, Q.B., K.C., L.Z. and Y.Z.; formal analysis, P.W.; investigation, P.W.; resources, Q.B.; data acquisition, P.W.; writing—original draft preparation, P.W.; writing—review and editing, Q.B. and K.C.; visualization, P.W.; supervision, Q.B. and K.C.; project administration, Q.B.; funding acquisition, Q.B. All authors have read and agreed to the published version of the manuscript.

Funding: This research work was supported by the National Natural Science Foundation of China (grant no. 52075129).

Institutional Review Board Statement: Not applicable.

Informed Consent Statement: Not applicable.

Data Availability Statement: The datasets used or analyzed during the current study are available from the corresponding author upon reasonable request.

Conflicts of Interest: The authors declare no conflict of interest.

References

1. Shang, Z.; Gao, D.; Jiang, Z.; Lu, Y. Towards less energy intensive heavy-duty machine tools: Power consumption characteristics and energy-saving strategies. *Energy* **2019**, *178*, 263–274. [CrossRef]
2. Li, C.; Li, L.; Tang, Y.; Zhu, Y.; Li, L. A comprehensive approach to parameters optimization of energy-aware CNC milling. *J. Intell. Manuf.* **2019**, *30*, 123–138. [CrossRef]
3. Hu, L.; Tang, R.; Cai, W.; Feng, Y.; Ma, X. Optimisation of cutting parameters for improving energy efficiency in machining process. *Robot. Comp. Int. Manuf.* **2019**, *59*, 406–416. [CrossRef]
4. Feng, C.; Chen, X.; Zhang, J.; Huang, Y.; Qu, Z. Minimizing the energy consumption of hole machining integrating the optimization of tool path and cutting parameters on CNC machines. *Int. J. Adv. Manuf. Technol.* **2022**, *121*, 215–228. [CrossRef]
5. Yan, J.; Li, L. Multi-objective optimization of milling parameters—The trade-offs between energy, production rate and cutting quality. *J. Clean. Prod.* **2013**, *52*, 462–471. [CrossRef]
6. Zhang, X.; Yu, T.; Dai, Y.; Qu, S.; Zhao, J. Energy consumption considering tool wear and optimization of cutting parameters in micro milling process. *Int. J Mech. Sci.* **2020**, *178*, 105628. [CrossRef]
7. Wang, W.; Tian, G.; Chen, M.; Tao, F.; Zhang, C.; AI-Ahmari, A.; Li, Z.; Jiang, Z. Dual-objective program and improved artificial bee colony for the optimization of energy-conscious milling parameters subject to multiple constraints. *J. Clean. Prod.* **2020**, *245*, 118714. [CrossRef]
8. Zhong, Q.; Tang, R.; Peng, T. Decision rules for energy consumption minimization during material removal process in turning. *J. Clean. Prod.* **2017**, *140*, 1819–1827. [CrossRef]
9. Li, L.; Yan, J.; Xing, Z. Energy requirements evaluation of milling machines based on thermal equilibrium and empirical modelling. *J. Clean. Prod.* **2013**, *52*, 113–121. [CrossRef]
10. Velchev, S.; Kolev, I.; Ivanov, K.; Gechevski, S. Empirical models for specific energy consumption and optimization of cutting parameters for minimizing energy consumption during turning. *J. Clean. Prod.* **2014**, *80*, 139–149. [CrossRef]
11. Gutowski, T.G.; Branham, M.S.; Dahmus, J.B.; Jones, A.J.; Thiriez, A. Thermodynamic analysis of resources used in manufacturing processes. *Environ. Sci. Technol.* **2009**, *43*, 1584–1590. [CrossRef] [PubMed]
12. Jiang, Z.; Gao, D.; Lu, Y.; Liu, X. Optimization of Cutting Parameters for Trade-off Among Carbon Emissions, Surface Roughness, and Processing Time. *Chin. J. Mech. Eng* **2019**, *32*, 94. [CrossRef]
13. Liu, Z.Y.; Sealy, M.P.; Guo, Y.B.; Liu, Z.Q. Energy Consumption Characteristics in Finish Hard Milling of Tool Steels. *Procedia Manuf.* **2015**, *1*, 477–486. [CrossRef]
14. Deng, Z.; Zhang, H.; Fu, Y.; Wan, L.; Liu, W. Optimization of process parameters for minimum energy consumption based on cutting specific energy consumption. *J. Clean. Prod.* **2017**, *166*, 1407–1414. [CrossRef]
15. Lv, J.; Tang, R.; Jia, S.; Liu, Y. Experimental study on energy consumption of computer numerical control machine tools. *J. Clean. Prod.* **2016**, *122*, 3864–3874. [CrossRef]
16. Lv, L.; Deng, Z.; Yan, C.; Liu, T.; Wan, L.; Gu, Q. Modelling and analysis for processing energy consumption of mechanism and data integrated machine tool. *Int. J. Prod. Res.* **2020**, *58*, 7078–7093. [CrossRef]
17. de Carvalho, H.M.B.; de Gomes Oliveira, J.; Schmidt, M.A.; Brandão, V.L.C. Vibration Analysis and Energy Efficiency in Interrupted Face Milling Processes. *Procedia CIRP* **2015**, *29*, 245–250. [CrossRef]
18. Pervaiz, S.; Deiab, I.; Rashid, A.; Nicolescu, M. Prediction of energy consumption and environmental implications for turning operation using finite element analysis. *Proc. Inst. Mech. Eng. Part B J. Eng. Manuf.* **2015**, *229*, 1925–1932. [CrossRef]
19. Shi, K.N.; Zhang, D.H.; Liu, N.; Wang, S.B.; Ren, J.X.; Wang, S.L. A novel energy consumption model for milling process considering tool wear progression. *J. Clean. Prod.* **2018**, *184*, 152–159. [CrossRef]
20. Li, C.; Tang, Y.; Cui, L.; Li, P. A quantitative approach to analyze carbon emissions of CNC-based machining systems. *J. Intell. Manuf.* **2015**, *26*, 911–922. [CrossRef]
21. Tian, C.; Zhou, G.; Lu, F.; Chen, Z.; Zou, L. An integrated multi-objective optimization approach to determine the optimal feature processing sequence and cutting parameters for carbon emissions savings of CNC machining. *Int. J. Comput. Integr. Manuf.* **2020**, *33*, 609–625. [CrossRef]
22. Zhou, L.; Li, J.; Li, F.; Meng, Q.; Li, J.; Xu, X. Energy consumption model and energy efficiency of machine tools: A comprehensive literature review. *J. Clean. Prod.* **2016**, *112*, 3721–3734. [CrossRef]
23. Jamil, M.; Zhao, W.; He, N.; Gupta, M.K.; Sarikaya, M.; Khan, A.M.; Sanjay, M.R.; Siengchin, S.; Pimenov, Y.D. Sustainable milling of Ti–6Al–4V: A trade-off between energy efficiency, carbon emissions and machining characteristics under MQL and cryogenic environment. *J. Clean. Prod.* **2021**, *281*, 125374. [CrossRef]
24. Singh, R.; Dureja, J.S.; Dogra, M.; Gupta, M.K.; Jamil, M.; Mia, M. Evaluating the sustainability pillars of energy and environment considering carbon emissions under machining ofTi-3Al-2.5 V. *Sustain. Energy Technol. Assess.* **2020**, *42*, 100806. [CrossRef]
25. Turhan, M.H.; Tseng, G.W.G.; Erkorkmaz, K.; Fidan, B. Dynamic model identification for CNC machine tool feed drives from in-process signals for virtual process planning. *Mechatronics* **2020**, *72*, 102445. [CrossRef]
26. Zheng, W.; Zhou, M.; Zhou, L. Influence of process parameters on surface topography in ultrasonic vibration- assisted end grinding of SiCp/Al composites. *Int. J. Adv. Manuf. Technol.* **2017**, *91*, 2347–2358. [CrossRef]
27. Liu, J. Research on Material Removal and Parameter Optimization in Micro Milling SICp/AL Composites. Ph.D. Thesis, Harbin Institute of Technology, Harbin, China, 2018; pp. 75–77. (In Chinese)

28. Chaudhari, P.; Thakur, K.A.; Kumar, R.; Banerjee, N.; Kumar, A. Comparison of NSGA-III with NSGA-II for multi objective optimization of adiabatic styrene reactor. *Mater. Today Proc.* **2022**, *57*, 1509–1514. [CrossRef]
29. Tavana, M.; Li, Z.; Mobin, M.; Komaki, M.; Teymourian, E. Multi-objective control chart design optimization using NSGA-III and MOPSO enhanced with DEA and TOPSIS. *Expert Syst. Appl.* **2016**, *50*, 17–39. [CrossRef]
30. Selvakumar, S.; Arulshri, K.P.; Padmanaban, K.P.; Sasikumar, K.S.K. Design and optimization of machining fixture layout using ANN and DOE. *Int. J. Adv. Manuf. Technol.* **2013**, *65*, 1573–1586. [CrossRef]
31. Wang, P.; Bai, Q.; Cheng, K.; Zhao, L.; Ding, H.; Zhang, Y. Machinability analysis of micro-milling thin-walled Ti-6Al-4V micro parts under dry, lubrication, and chatter mitigation conditions. *Proc. Inst. Mech. Eng. Part B J. Eng. Manuf.* **2023**, 1–12. [CrossRef]

applied sciences

MDPI

Article

Investigation on Tool Path Planning Algorithm of Propeller Blade Double-Sided Collaborative Machining

Rui Wang [1,*], Yuhao Ge [1] and Xiangyu Guo [2]

[1] Weihai Key Laboratory of Intelligent Operation and Maintenance, Harbin Institute of Technology, No. 2 Wenhuaxi Road, Weihai 264200, China; bfdgtx@163.com

[2] School of Mechatronics Engineering, Harbin Institute of Technology, No. 92 West Dazhi Street, Harbin 150001, China; 19b908100@stu.hit.edu.cn

* Correspondence: wrhit@163.com

Abstract: The concomitant vibration and deformation produced by propeller blades in single-sided machining seriously affect the surface machining precision. Double-sided symmetrical machining can improve system rigidity through mutual shoring on both sides which abates the concomitant vibration and deformation. However, the actual double-sided symmetrical machining cannot be applied to blade machining due to its shape complexity. The double-sided collaborative machining method combining symmetrical machining and staggered machining is devised in this paper, and its tool path planning algorithm is investigated. Firstly, the algorithm achieves smooth fitting and correspondence of bilateral cutter position points through double-curve interpolation and position data alignment. Secondly, the blade surface is divided into four regions by two partition parameters: tip region, internal region, variable region, and edge region. Then, the conversion between symmetrical machining and staggered machining is completed through the Sigmoid deformation curve in the variable region. Finally, the feasibility and superiority of double-sided collaborative machining are verified through machining experiments.

Keywords: propeller; double-sided machining; tool path planning; double-curve interpolation; position data alignment; Sigmoid curve

check for updates

Citation: Wang, R.; Ge, Y.; Guo, X. Investigation on Tool Path Planning Algorithm of Propeller Blade Double-Sided Collaborative Machining. *Appl. Sci.* **2023**, *13*, 6529. https://doi.org/10.3390/app13116529

Academic Editors: Rui Manuel Leal, Gustavo H. S. F. L. Carvalho and Ivan Galvão

Received: 8 April 2023
Revised: 22 May 2023
Accepted: 23 May 2023
Published: 27 May 2023

1. Introduction

As the main component of the propulsion device, propellers importantly affect the propulsion ability of ships. Machining defects such as scratches, protrusions, and burrs on the blades will increase the frictional resistance between blades and water flow during the propulsion process which reduces fuel conversion and ship speed. Consequently, ameliorating the machining quality of blades is of great significance to improve the propulsion ability of ships. Large propeller blades are mostly manufactured by heavy-duty CNC machine tools, such as the CNC seven-axis five-link propeller machining heavy-duty turning and milling machine tool CKX5680 led by Wuhan Heavy Duty Machine Tool Group Corporation [1], the CNC seven-axis six-link propeller machining heavy-duty machine tool VTM11000 led by Zhongchuan Heavy Duty Machine Tool Corporation Limited, and the CNC machine tool HPMC-110 of HNK Machine Tool [2]. However, there are two shortcomings with heavy-duty CNC machine tools: (1) After completing the single-sided machining of the propeller, it is necessary to turn the propeller over through a special flipping mechanism, and the opposite side of the propeller can only be machined after the secondary clamping. The flipping process will cause a certain degree of damage to the machined surface, and secondary clamping will result in positioning deviation on both sides. (2) The blade is in a cantilever state during machining, and the current machining auxiliary shoring cannot completely neutralize the concomitant vibration and deformation caused by the cutting force; consequently, the improvement of surface machining accuracy is limited.

This investigation solves the shortcomings in traditional single-sided machining of propellers by means of double-sided machining. In recent years, academia has produced a profound study on the theory of double-sided machining [3–15]. Mohanad Kadhim Mejbel, Isam Tareq Abdullah, et al. have studied the simultaneous machining of thin-walled surfaces with double-end cutter milling, which greatly improves the flatness and straightness of the workpiece surface [3]. A novel double-sided milling strategy of thin-walled parts based on dual collaborative parallel kinematic machines was designed by Rao Fu, Patrick Curley, et al. The double-sided milling strategy doubles the material removal rate while eliminating the cumbersome in-process steps used in conventional machining [4]; Tatsuya Mori, Tomoki Hiramatsu, Eiji Shamoto, et al. presented a new method to finish flexible plates with high accuracy and high productivity, in which both surfaces are finished simultaneously with synchronization so that the thrust forces are canceled out on both sides [5–7]. Hao Jinming, Zhao Yong, et al. introduced a newly designed mirror machining device and analyzed the stiffness of the support mechanism and the workpiece. The distribution of the synthetical stiffness is obtained from the stiffness model and improved by adding redundant actuation. The results show that the stiffness of the support mechanism and the minimum synthetical stiffness are significantly improved [8]. Song Dongdong, Xue Fei, et al. proposed a new path smoothing algorithm based on the geometric constraints to guarantee the smoothness of twin-tool paths and the synchronization of the two cutters for the nine-axis twin-tool milling of blade profiles. The results of simulation and experiments show that the two surfaces of the blade can be simultaneously and entirely machined, and the contour errors do not exceed the tolerance range [9]. Lu, Fang, et al. used a newly developed ISF machine and investigated the process of double-side incremental forming [10]. The interaction method has a certain reference significance for double-sided machining. Bo Qile proposed a thin-walled parts mirror milling stability monitoring technology and the developed mirror milling system and successfully realized the efficient and stable mirror milling of large thin-walled parts [11]. Gangaram Mandaloi, Aniket Ramnath Nagargoje, et al. presented a novel double-sided deformation machining (DSDM) technique, and the machining and incremental forming operations are combined to utilize the strengths of the two processes to produce monolithic components with thin structures using a single setup [12]. Our team has conducted partial research on the double-sided machining technology of propeller blades using the XYZ-3RPS hybrid machine tool and achieved machining control of the cutters on both sides [13,14]. However, investigation on double-sided machining mostly focuses on uniform thin-walled workpieces currently, which have simple and symmetrical shapes. Double-sided symmetrical machining of these workpieces improves system rigidity through mutual shoring of two side cutters, thereby greatly improving surface machining accuracy and efficiency. However, most propeller blades are asymmetric complex workpieces, and the existing double-sided symmetrical machining methods are not suitable for blade machining. Erhan Budak, Alptunc Comak, et al. found that although double-sided milling has significant advantages in machining efficiency, parameter selection is quite challenging due to the dynamic interaction between the tools [15]. Similarly, the interaction and coordination design between the tools is also significant in propeller blade double-sided machining due to the characteristics of thick middle and thin edge. Considering the advantages of double-sided symmetrical machining, this paper presents a propeller blade double-sided collaborative machining method combining double-sided symmetrical machining with double-sided staggered machining and its tool path planning algorithm.

2. Overall Design of Double-Sided Machining Tool Path

It is inevitable to leave tool marks on the surface of propeller blades during machining. If the direction of the surface machining tool marks can be kept as consistent as possible with the running streamline of the propeller surface, the adverse impact of their roughness on flow stability will be greatly reduced [16,17]. However, the shapes of the propeller blade's surface streamline with different types and are not the same. The direction of the running streamline of the general propeller surface is similar to the projection curve of the propeller concentric circles on its surface. Thus, the method used in this paper for generating a machining tool path on the surface of a general propeller is as follows: (1) Make a series of concentric circles for the target propeller through a setting radius increment ΔR until the concentric circles' radius is greater than or equal to the propeller radius. (2) Project the concentric circles onto the blade surface to obtain a set of contour curves. (3) The contour curves are divided into two parts, the leading edge and trailing edge of the blade, which are used as tool paths on both sides. (4) Connect adjacent tool paths to form a reciprocating machining tool path, as shown in Figure 1. The radius increment ΔR is generally less than or equal to one-quarter of the radius of the ball nose end mill in the generation method. In the reciprocating machining, the tool paths generated from contours curves are called machining-tool-path, and the connection of the machining-tool-path is called turning-tool-path.

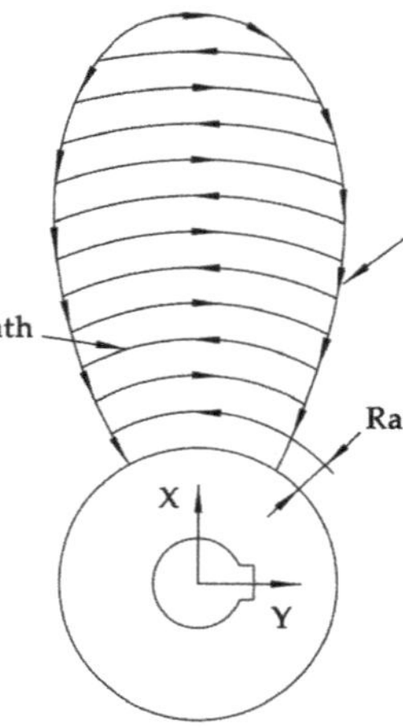

Figure 1. Reciprocating tool paths of the propeller blade.

During the propeller blade double-sided machining, there are four force-bearing positions in the machining system: the clamping support surface and tool contact points on both sides, and machining torques inevitably occur between force-bearing positions. The machining torque generated in the machining system mainly includes two types: the overturning torque between the clamping support surface and the tool contact point, and the torsional torque between the tool contact points on both sides. The size of the overturning torque is positively correlated with the distance between the tool contact and the clamping support surface. The size of the torsional torque is positively correlated with the relative distance between the tool contacts on both sides. Both machining torques can cause concomitant vibration and deformation, thereby reducing the machining accuracy of the blade surface. Therefore, it is necessary to design an interaction and coordination mode between the tools to improve system rigidity or reduce the size of the lever arm.

The milling cutters on both sides support each other to neutralize the cutting force and improve system rigidity in the double-sided symmetrical machining of uniform thin-walled workpieces, and the coincidence of the cutter axes makes the torsional moment basically zero. However, most propeller blades are asymmetric complex workpieces, which is not suitable for double-sided symmetrical machining in the general sense. Based on the theory of double-sided symmetric machining, the relevant definitions of "symmetric machining" of propeller blades in this paper are as follows: (1) Each of the two machining-tool-paths obtained by dividing the same contour curve is called a pair of symmetrical-tool-paths of the propeller blade. (2) Insert the same number of cutter position points evenly on a pair of symmetrical-tool-paths through position data alignment. Each pair of cutter position points is called a pair of symmetrical-cutter-position-points of the propeller blade, as shown in Figure 2. (3) The process of overlapping the cutter axes on both sides with the line connecting the symmetrical-cutter-position-points and machining in the direction of the reciprocating machining tool path is called symmetrical machining of the propeller blades. Symmetrical machining neutralizes the axial cutting force and improves system rigidity through the axial mutual shoring of both side cutters, which can greatly reduce the adverse impact of overturning and torsional torques. Due to the characteristics of the thick middle and thin edge of the blade, symmetrical machining at the blade edge is prone to collision accidents with low security. Therefore, it is necessary to stagger the cutters on both sides and convert to staggered machining when double-sided machining to the propeller blade edge. In this study, a symmetrical XYZ-3RPS double-sided hybrid machine tool was used to complete the validation of the double-sided collaborative machining of propeller blades, as shown in Figure 3.

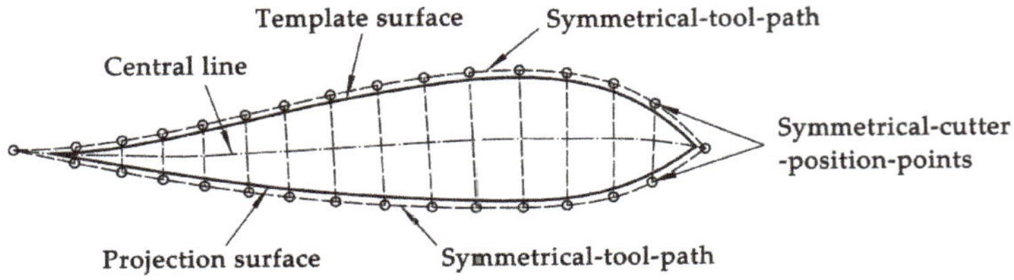

Figure 2. Symmetrical-cutter-position-points of symmetrical-tool-paths.

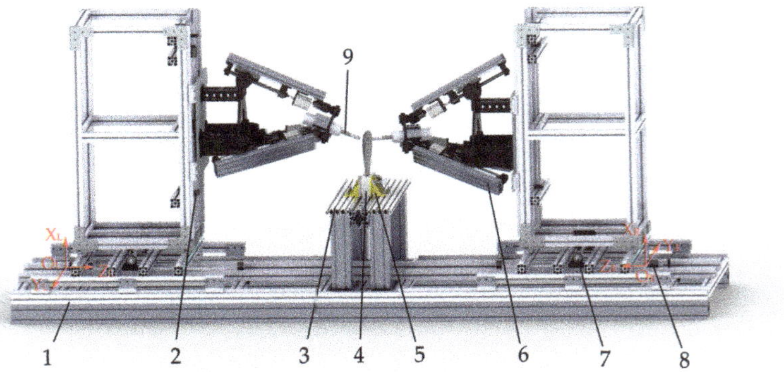

Figure 3. Symmetrical XYZ-3RPS double-sided hybrid machine tool: (1) machine base; (2) X-axis lead screw; (3) workbench; (4) propeller blade workpiece; (5) clamping support; (6) 3RPS parallel mechanism; (7) Y-axis lead screw; (8) Z-axis lead screw; and (9) ball nose end mill.

3. Cutter Position Points Interpolation and Data Alignment on Both Sides

The tool path curves between adjacent tool position points are fitted by piecewise polynomial interpolation during machining, but unrelated fluctuations to the ideal curve may occur in a series of cutter position points fitting processes. The causes of unrelated fluctuations are generally uneven cutter position points or excessive spacing, which can also lead to discontinuity in the cutter-orientation transformation. Therefore, in this paper, a cubic spline curve is applied to fit the cutter position points path and cutter tail points path and perform equidistant interpolation on the fitting curve after determining the number of interpolation steps, as shown in Figure 4. In the machine coordinate system, the cutter-orientation $(i_{k,n}, j_{k,n}, k_{k,n})$ of the interpolation point is obtained by using the interpolation cutter position point coordinates $(x_{k,n}, y_{k,n}, z_{k,n})$ and its corresponding interpolation cutter tail point coordinates $(x_{k,n}{}', y_{k,n}{}', z_{k,n}{}')$.

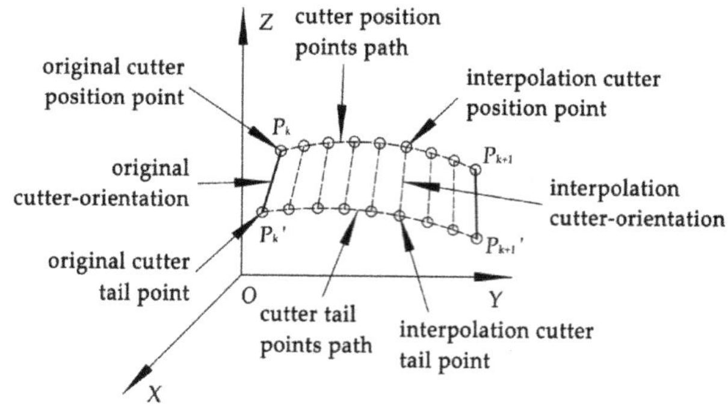

Figure 4. Interpolation of cutter position points coordinates and orientation.

The number of original cutter position points on both sides is inconsistent due to the complex shapes of the blade curved surfaces, and it is impossible to achieve double-sided symmetrical machining. Therefore, it is necessary to achieve the correspondence of cutter position points on both sides through data alignment processing. According to the number of original cutter position points on both sides of the blade, select the side with fewer cutter position points as the projection surface and the other side as the template surface. Firstly, interpolate the cutter position points on the template surface. Each tool path between two original cutter position points is regarded as an interpolation interval, and it is assumed that there are K interpolation intervals in a machining-tool-path. In the machine coordinate system, the interpolation interval starting point coordinates $p_k = (x_k, y_k, z_k)^T$ and orientations $w_k = (i_k, j_k, k_k)^T$, and the ending point coordinates $p_{k+1} = (x_{k+1}, y_{k+1}, z_{k+1})^T$ and orientations $w_{k+1} = (i_{k+1}, j_{k+1}, k_{k+1})^T$, where $k = 1, 2, \cdots, K$. Thereby, the starting points and ending points of the cutter tail point interpolation intervals are obtained as follows:

$$p_k' = \begin{bmatrix} x_k' \\ y_k' \\ z_k' \end{bmatrix} = \begin{bmatrix} x_k \\ y_k \\ z_k \end{bmatrix} + d_T \begin{bmatrix} i_k \\ j_k \\ k_k \end{bmatrix}, \tag{1}$$

$$p_{k+1}' = \begin{bmatrix} x_{k+1}' \\ y_{k+1}' \\ z_{k+1}' \end{bmatrix} = \begin{bmatrix} x_{k+1} \\ y_{k+1} \\ z_{k+1} \end{bmatrix} + d_T \begin{bmatrix} i_{k+1} \\ j_{k+1} \\ k_{k+1} \end{bmatrix}, \tag{2}$$

where d_T represents the tool length.

A single machining-tool-path is a spatial curve, and the coordinates of the Y-axis monotonically increase or decrease in the machine coordinate system. Therefore, the projection curve of a spatial curve can be fitted on the Y-X plane and the Y-Z plane, and the spatial curve of a single machining tool path can be expressed in the form of parametric equations with $y = t$:

$$x = f_1(t), \tag{3}$$

$$z = g_1(t), \tag{4}$$

The path length of each interpolation interval of a single machining-tool-path can be obtained as follows:

$$s_k = \int_{t_k}^{t_{k+1}} \sqrt{\left(f_1^{(1)}(t)\right)^2 + \left(g_1^{(1)}(t)\right)^2 + 1}, \tag{5}$$

Similarly, the path length of each interpolation interval of the cutter tail points can be obtained:

$$s_k' = \int_{t_k'}^{t_{k+1}'} \sqrt{\left(f_2^{(1)}(t)\right)^2 + \left(g_2^{(1)}(t)\right)^2 + 1}, \tag{6}$$

Calculate the number of interpolation steps N according to the set interpolation step size Δd:

$$N = \max\left(\frac{s_k}{\Delta d}+1, \frac{s_k{}'}{\Delta d}+1\right), \tag{7}$$

The path length sequence $[s_{k,1}, s_{k,2}, \cdots, s_{k,N}]$ from the interpolation cutter position points to the interpolation interval starting point can be obtained by using the interpolation steps N:

$$s_{k,n} = \frac{ns_k}{N}, n = 1,2,3,\cdots,N, \tag{8}$$

Similarly, the path length sequence $[s_{k,1}{}', s_{k,2}{}', \cdots, s_{k,N}{}']$ of cutter tail interpolation points can be obtained:

$$s_{k,n}{}' = \frac{ns_k{}'}{N}, n = 1,2,3,\cdots,N, \tag{9}$$

Then, iteratively calculate the equation parameter sequence $[t_{k,1}, t_{k,2}, \cdots, t_{k,N}]$ and $[t_{k,1}{}', t_{k,2}{}', \cdots, t_{k,N}{}']$ through numerical method:

$$s_{k,n} = \int_{t_k}^{t_k+i\Delta t} \sqrt{\left(f_1{}^{(1)}(t)\right)^2 + \left(g_1{}^{(1)}(t)\right)^2 + 1}, \tag{10}$$

$$s_{k,n}{}' = \int_{t_k{}'}^{t_k{}'+j\Delta t} \sqrt{\left(f_2{}^{(1)}(t)\right)^2 + \left(g_2{}^{(1)}(t)\right)^2 + 1}, \tag{11}$$

$$t_{k,n} = t_k + i\Delta t, \tag{12}$$

$$t_{k,n}{}' = t_k{}' + j\Delta t, \tag{13}$$

where i and j represent the number of iterations: $i = 1,2,3,\cdots$, $j = 1,2,3,\cdots$; Δt is the iteration step length.

The interpolation cutter position point coordinates $p_{k,n} = (x_{k,n}, y_{k,n}, z_{k,n})$ can be obtained by substituting the equation parameter sequence $[t_{k,1}, t_{k,2}, \cdots, t_{k,N}]$ into Equations (3) and (4). Similarly, the interpolation cutter tail point coordinates $p_{k,n}{}' = (x_{k,n}{}', y_{k,n}{}', z_{k,n}{}')$ can be obtained. Finally, cutter-orientation interpolation is performed through $p_{k,n}$ and $p_{k,n}{}'$:

$$w_{k,n} = (i_{k,n}, j_{k,n}, k_{k,n}) = \frac{p_{k,n} - p_{k,n}{}'}{\Delta l}, \tag{14}$$

$$\Delta l = \sqrt{(x_{k,n} - x_{k,n}{}')^2 + (y_{k,n} - y_{k,n}{}')^2 + (z_{k,n} - z_{k,n}{}')^2}, \tag{15}$$

After interpolating the cutter position points of the template surface, simply align the interpolation points on the template surface to the projected surface. The uniformity of cutter position points on both sides should be consistent after position data alignment, as shown in Figure 2. The path length sequence $[S_1, S_2, \cdots, S_K]$ from the original cutter position points of a single machining-tool-path to the starting point of this tool path before interpolation is as follows:

$$S_k = \sum_1^k s_\kappa, \tag{16}$$

where: $k = 1,2,\cdots,K$; $\kappa = 1,2,\cdots,k$.

The path length S of a single machining-tool-path is as follows:

$$S = \sum_1^K s_k, \tag{17}$$

The path length sequence $[S_{k,1}, S_{k,2}, \cdots, S_{k,N}]$ from the interpolation cutter position points to the starting point of this tool path in each interpolation interval of a single machining-tool-path on the template plane can be calculated using the path length sequence $[s_{k,1}, s_{k,2}, \cdots, s_{k,N}]$ of the interpolation point, where $S_0 = 0$:

$$[S_{k,1}, S_{k,2}, \cdots, S_{k,N}] = [s_{k,1}, s_{k,2}, \cdots, s_{k,N}] + S_{k-1}, \tag{18}$$

Based on the ratio of the sequence $[S_1, S_2, \cdots, S_K]$ and $[S_{k,1}, S_{k,2}, \cdots, S_{k,N}]$ to the length of a single machining-tool-path, the path length ratio sequence $[\xi_2, \xi_3, \cdots, \xi_{D-1}]$ from all cutter position points except the starting and ending points to the starting point of the machining-tool-path on the template surface can be obtained after interpolation, where D is the number of cutter position points.

Then, the original cutter position points are used to fit the machining-tool-path curve of the projection surface. Referring to Equations (3) to (6), assuming that the Y-axis coordinates of the starting and ending points of a single machining-tool-path on the projection surface are t_1^c and t_D^c, the tool path length is as follows:

$$S^c = \int_{t_1^c}^{t_D^c} \sqrt{\left(f_3{}^{(1)}(t)\right)^2 + \left(g_3{}^{(1)}(t)\right)^2 + 1}, \tag{19}$$

Multiply the tool path length S^c by the elements in the path length ratio sequence $[\xi_2, \xi_3, \cdots, \xi_{D-1}]$, and iterate through the numerical method to calculate the equation parameter sequence $[t_2^c, t_3^c, \cdots, t_{D-1}^c]$:

$$S^c \cdot \xi_n = \int_{t_1^c}^{t_1^c + b\Delta t} \sqrt{\left(f_3{}^{(1)}(t)\right)^2 + \left(g_3{}^{(1)}(t)\right)^2 + 1}, \tag{20}$$

$$t_n^c = t_1^c + b\Delta t, \tag{21}$$

where b represents the number of iterations: $b = 1, 2, 3, \cdots$; Δt is the iteration step length; $n = 2, 3, \cdots, D-1$.

Finally, the corresponding cutter position point coordinates can be obtained by substituting the equation parameter sequence into the tool path parameter equation of the projection surface, and the cutter-orientations remain consistent.

4. Propeller Blade Region Division

First, define two region division parameters before propeller blade region division: minimum safety distance h_{safe}, and maximum staggered ratio K_{max}. The minimum safety distance refers to the minimum safety machining distance between the cutter tips on both sides during double-sided machining. The parameter size depends on the radius of the ball nose end mill in double-sided machining. The safety of double-sided machining can be ensured when the spacing between cutter tips on both sides is greater than or equal to the cutter radius; thus, the parameter size of the minimum safety distance is generally set to 1–1.5 times the radius r_{mill}. The staggered ratio refers to the length proportion of the staggered machining tool path in a pair of symmetrical-tool-paths whose range is twice the range that does not meet the division parameter h_{safe} to the symmetrical-tool-paths, and the maximum staggered ratio K_{max} is the parameter that constrains the length proportion. To ensure that the propeller blades are machined symmetrically on both sides to the maximum extent, the maximum staggered ratio K_{max} is generally set to 0.1–0.2. Through two region division parameters, the blade surface is divided into four regions in this paper: tip region, internal region, variable region, and edge region. The division rules for the four regions are shown in Table 1.

Table 1. Rules for propeller blade region division.

Division Rules of K_{max}	Division Rules of h_{safe}	Corresponding Regions
staggered ratio $\geq K_{max}$	—	tip region
staggered ratio $< K_{max}$	the spacing between cutter tips on both sides $\leq h_{safe}$	edge region
	the spacing between cutter tips on both sides $> h_{safe}$ (Path length equal to edge region)	variable region
	the spacing between cutter tips on both sides $> h_{safe}$	internal region

The tip region, which does not meet the division parameter K_{max}, is generally thinner due to being away from the blade root. It is unable to guarantee the safety of double-sided symmetrical machining in this region; thus, single-sided alternating machining is implemented. Edge region, variable region, and internal region are obtained by dividing the machining-tool-paths meeting the division parameter K_{max}. The edge region is the portion of each pair of symmetrical-tool-paths that does not meet the division parameter h_{safe}. Due to problems such as insufficient spacing and the intersection of tool paths, it is also unable to guarantee the safety of double-sided symmetrical machining in this region; thus, double-sided staggered machining is used instead. The variable region is a transition region that connects the edge region and internal region, whose length is equal to the edge region. In this region, conversion between symmetrical machining and staggered machining is achieved by the interpolation of machining cutter points through the sigmoid deformation curve. The internal area meets the division parameter h_{safe} and implements double-sided symmetrical machining. The division of the propeller blade surface is shown in Figure 5.

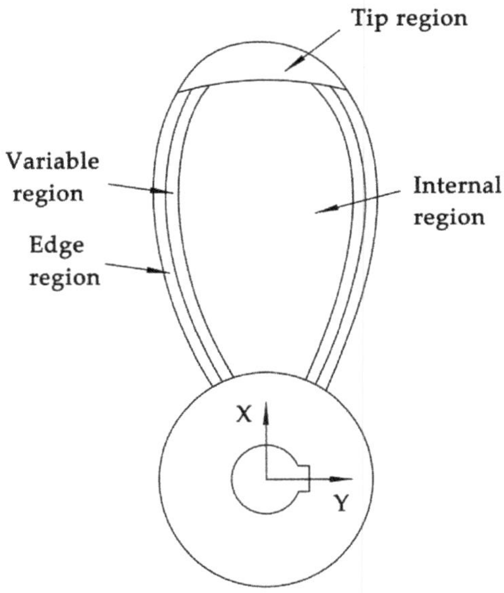

Figure 5. The division of the propeller blade surface.

5. Tool Path Planning Design of Double-Sided Collaborative Machining

The cutters on both sides are defined as the leading cutter and subsequent cutter, respectively, in the double-sided collaborative machining. In each pair of symmetrical-tool-paths, the cutter that finishes machining first is called the leading cutter, and the other is called the subsequent cutter. The blade surfaces corresponding to the cutters on both sides are called the leading surface and subsequent surface, respectively. In addition, the double-sided collaborative machining also requires six boundary cutter position points to describe the tool path planning process, as shown in Figure 6: speed varied starting point of subsequent surface S_{vs}, staggered starting point of leading surface S_{sl}, speed varied ending point of subsequent surface E_{vs}, speed varied starting point of leading surface S_{vl}, staggered starting point of subsequent surface E_{ss}, speed varied ending point of subsequent leading E_{vl}.

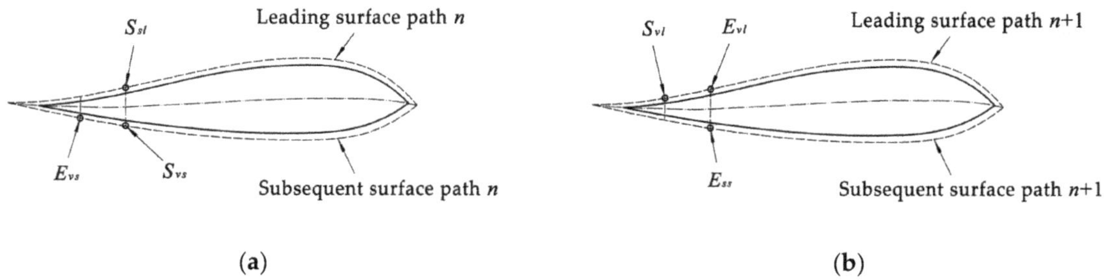

Figure 6. (**a**) Boundary cutter position points S_{vs}, S_{sl}, E_{vs} of end stage on path n. (**b**) Boundary cutter position points S_{vl}, E_{ss}, E_{vl} of start stage on path n + 1.

S_{vs} is the boundary point between the internal region and variable region at the end stage of each symmetrical tool path on the subsequent surface. S_{sl} is the corresponding point of S_{vs} and the starting point of staggered machining. E_{vs} is the boundary point between the variable region and edge region at the end stage of each symmetrical tool path on the subsequent surface. S_{vl} is the boundary point between the edge region and variable region at the start stage of each symmetrical tool path on the leading surface. E_{vl} is the boundary point between the variable region and internal region at the start stage of each symmetrical tool path on the leading surface. E_{ss} is the corresponding point of E_{vl} and the ending point of staggered machining. The design of the tool path planning process for double-sided collaborative machining through the above six boundary tool points is as follows:

1. The leading cutter machines the tip region on one side along the machining-tool-path of the leading surface, and the subsequent cutter waits at the zero position;
2. Assuming that the last machining-tool-path of the tip region is path $n - 1$, the leading cutter lifts at S_{vl} on path n and waits, while the subsequent cutter starts machining the tip region on one side;
3. When the subsequent cutter machines at the corresponding cutter position point of S_{vl} on path n, the leading cutter enters and starts double-sided symmetrical machining, as shown in Figure 7a;
4. The subsequent cutter and leading cutter machines to S_{vs} and S_{sl}, respectively, on path n, and the subsequent cutter enters the S-type variable speed stage through the sigmoid deformation curve and starts double-sided staggered machining, as shown in Figure 7b;
5. The leading cutter enters the S-type variable speed stage through the sigmoid deformation curve when it completes the machining of path n and machines to S_{vl} on path n + 1, and the subsequent cutter machines to E_{vs} on path n and exits the S-type variable speed stage at the same time, as shown in Figure 7c;
6. When the subsequent cutter completes the machining of path n and machines to E_{ss} on path n + 1, the leading cutter machines to E_{vl} on path n + 1 and exits the S-type variable speed stage, and implement double-sided symmetrical machining again, as shown in Figure 7d;
7. Repeat steps 4 to 6 until the leading cutter and the subsequent cutter return to the zero position.

During the single-sided machining of the tip region, the cutter-orientations coincide with the normal direction of the machining cutter position points. During the double-sided staggered machining, the cutter-orientations coincide with the line connecting the symmetrical-cutter-position-points before the staggered machining.

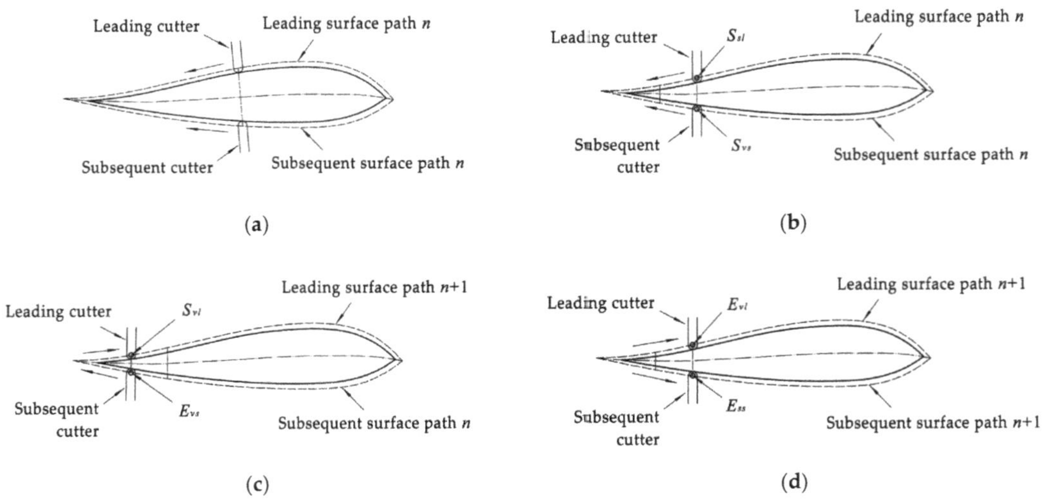

Figure 7. (**a**) Double-sided symmetrical machining of the internal region. (**b**) Start of double-sided staggered machining. (**c**) Meeting of cutters on both sides. (**d**) End of double-sided staggered machining.

6. Sigmoid Deformation Curve Interpolation

Compared with the existing variable speed methods, such as trapezoidal variable speed, exponential variable speed, trigonometric function variable speed, etc. [18], the S-type variable speed methods such as exponential variable speed and trigonometric function variable speed are stable and reliable, and it is suitable for frequent acceleration and deceleration. Therefore, the S-type variable speed is adopted as the variable speed method in the variable region. As a common S-type mathematical squashing function, the sigmoid function can squash the input between 0 and 1, which is beneficial for the deformation of the sigmoid function in research. In the tool path planning of double-sided collaborative machining, the sigmoid deformation curve is obtained by deforming the sigmoid function. and this curve is used to complete the conversion between double-sided symmetrical machining and double-sided staggered machining in the S-type variable speed stage.

As shown by the solid line in Figure 8, the sigmoid deformation curve includes the deceleration stage in the first half and the acceleration stage in the second half. A smooth and stable variable speed can be achieved by using the sigmoid deformation curve for interpolation in the variable region. Equation (22) is the mathematical expression of the sigmoid deformation curve.

$$\begin{cases} V(t) = \frac{K_V}{1+e^{K_a(t-K_T)}} + K, & t \in [0, 2K_T] \\ V(t) = \frac{K_V}{1+e^{K_a(3K_T-t)}} + K, & t \in [2K_T, 4K_T] \\ K = \frac{K_V e^{-K_a K_T}}{1+e^{-K_a K_T}} \end{cases}, \tag{22}$$

where K_T is time adjustment parameter that controls the range of the independent variable interval, which is equal to 0.25 times the variable speed time; K_V is the velocity adjustment parameter that controls the range of the dependent variable interval, which is equal to the initial velocity of the variable speed stage; K_a is acceleration adjustment parameter that controls the slope of the sigmoid deformation curve, which is solved by Equations (24) and (25); The constant term K represents the difference between the actual velocity and the initial and final velocity of the sigmoid deformation curve.

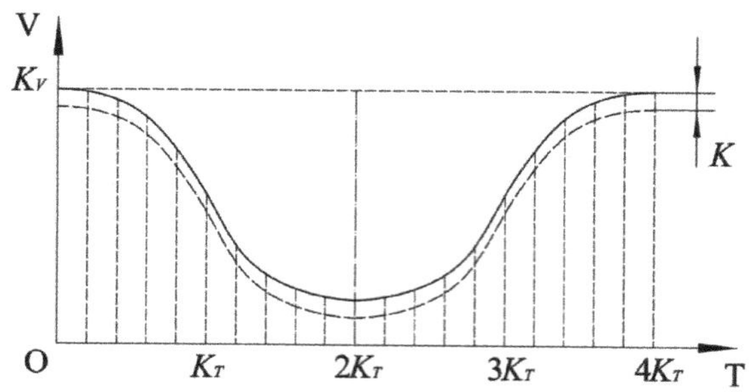

Figure 8. The sigmoid deformation curve.

The following displacement curve expression is obtained by integrating Equation (22) within the interval $[0,\ 4K_T]$:

$$\begin{cases} I(t) = -\frac{K_V}{K_a}\ln(1+e^{-K_a(t-K_T)}) + Kt, & t \in [0, 2K_T] \\ I(t) = \frac{K_V}{K_a}\ln(1+e^{-K_a(3K_T-t)}) + Kt, & t \in [2K_T, 4K_T] \\ K = \frac{K_V e^{-K_a K_T}}{1+e^{-K_a K_T}} \end{cases}, \tag{23}$$

Substitute the interval length L_c of the variable speed region into Equation (22) to simplify the monadic equation of acceleration adjustment parameter K_a:

$$2K_V \ln(1+e^{K_a K_T}) - (0.5L_c + K_V K_T)K_a + \frac{2K_V K_T}{1+e^{K_a K_T}} = 0, \tag{24}$$

Use the numerical method to calculate Equation (24) and obtain the initial acceleration adjustment parameter K_a. Due to the upper limit of the driving axis acceleration of the machine tool being $a_{\max}$, it can be inferred from the derivative of Equation (22) that the slope of the sigmoid deformation curve is the highest at $t = K_T$ and $t = 3K_T$, which the maximum acceleration value during the variable speed stage. Substitute $t = K_T$ and $t = 3K_T$ into the derivative expression of Equation (22), and use the upper limit of acceleration $a_{\max}$ to constrain K_a:

$$\frac{K_V K_a}{4} \le a_{\max}, \tag{25}$$

If the initial K_a satisfies Equation (25), substitute it as the final acceleration adjustment parameter into Equations (22) and (23) to obtain the expressions of the sigmoid deformation curve and displacement curve. If the initial K_a does not satisfy Equation (25), increase the variable speed time to reduce the maximum acceleration during the variable speed stage.

7. Experiment

The cutter position points obtained by the tool path planning algorithm are further processed by the post-processing algorithm to calculate the corresponding displacement data and driving speed data of each axis. Finally, the driving data are integrated into a machining file and imported into the motion control system of the symmetrical XYZ-3RPS double-sided hybrid machine tool. Single-sided machining and double-sided collaborative machining experiments were conducted on nylon material workpieces (350 × 200 × 50) using a 3 mm ball nose end mill, with an axial depth of 2 mm and the same machining tool path. As shown in Figure 3, the machine tool is divided into left and right sides, and partial double-sided collaborative driving data on the left side are shown in Tables 2 and 3. x, y, z, d_1, d_2, d_3 represent the driving displacement of the six driving screws of the XYZ-3RPS hybrid machine tool on the left side, while v_x, v_y, v_z, v_1, v_2, v_3 represent the driving speed of the six driving screws of the XYZ-3RPS hybrid machine tool on the left side. The double-sided collaborative machining process of propeller blades is shown in Figure 9.

Table 2. Partial time-displacement data on the left side.

t (ms)	x (mm)	y (mm)	z (mm)	d_1 (mm)	d_2 (mm)	d_3 (mm)
14,195	15.8333	89.7109	70.9832	0.9979	−5.5220	4.6788
14,204	15.8109	89.5776	71.0373	0.9966	−5.5141	4.6717
14,213	15.7902	89.4441	71.0917	0.9953	−5.5063	4.6646
...	...	...	...	...	...	...
867,903	100.8913	97.2651	66.6947	6.7645	−8.5218	2.1236
867,975	101.0091	98.4719	66.4462	6.7705	−8.5876	2.1882
868,047	101.1385	99.6754	66.2085	6.7772	−8.6534	2.2524

Table 3. Partial time-velocity data on the left side.

t (ms)	v_x (mm/ms)	v_y (mm/ms)	v_z (mm/ms)	v_1 (mm/ms)	v_2 (mm/ms)	v_3 (mm/ms)
14,195	0.00264	−0.01474	0.00599	−0.00016	0.00088	−0.00077
14,204	0.00237	−0.01483	0.00603	−0.00014	0.00088	−0.00079
14,213	0.00224	−0.01485	0.00604	−0.00013	0.00087	−0.00079
...	...	...	...	...	...	...
867,903	−0.00155	0.01675	−0.00349	0.00008	−0.00091	0.00090
867,975	−0.00171	0.01672	−0.00336	0.00009	−0.00091	0.00089
868,047	−0.00189	0.01681	−0.00331	0.00010	−0.00092	0.00089

Figure 9. The double-sided collaborative machining process of propeller blades.

With the help of a laser rangefinder and detection board installed on the blade, the deformation and vibration data of single-sided machining on the left side and double-sided collaborative machining are collected and integrated into this experiment. The comparison curves of deformation and vibration during the machining process are shown in Figures 10 and 11. The red curve represents the axial deformation of the blade, while the blue curve represents the axial vibration of the blade.

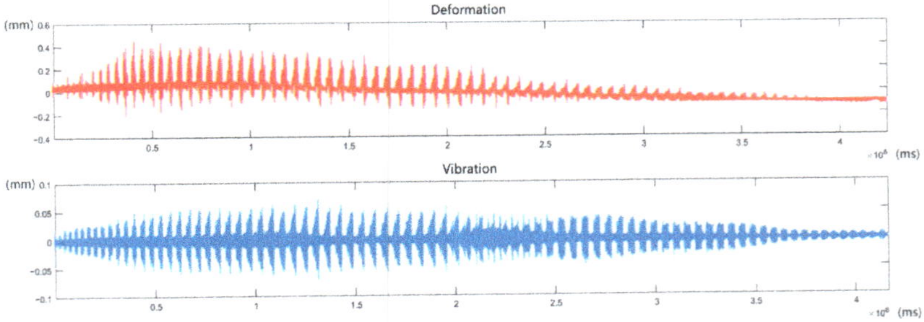

Figure 10. The deformation and vibration curves of single-sided machining.

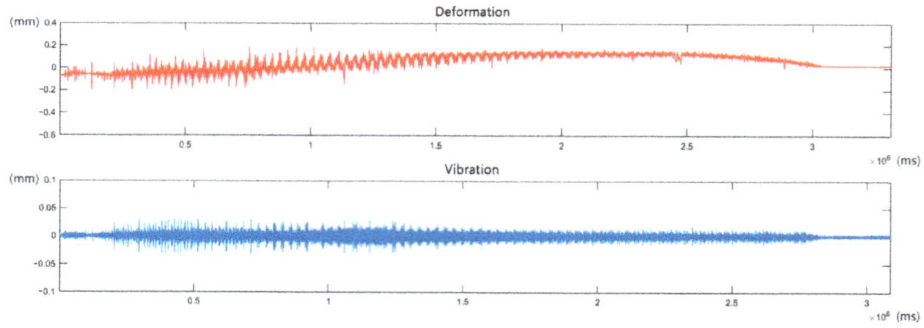

Figure 11. The deformation and vibration curves of double-sided collaborative machining.

Through the analysis of curve data, the comparison data of evaluation indicators can be obtained as shown in Table 4. It can be concluded that the deformation range of single-sided machining is between −0.209~0.456 mm, and the deformation range of double-sided collaborative machining is between −0.193~0.138 mm. By comparing the deformation range results, it can be seen that the double-sided collaborative machining method to some extent weakens the local influence of overturning torque and torsional torque on workpiece deformation during single-sided machining. Especially during each cycle of machining back and forth, the problem of machining deformation is effectively improved. However, the uneven force distribution on both sides of the machining area and the deformation resilience can still lead to significant deformation from the perspective of the entire machining process. Therefore, the process parameters of the double-sided collaborative machining method still need to be optimized in subsequent research. X_p, PP, rms, and CL represent the peak value of vibration, the peak-to-peak value of vibration, the root mean square value of vibration, and margin indicators, respectively. The vibration peak value and peak-to-peak value reflect the amplitude and range of the variation curve. The root mean square value of vibration reflects the strength and stability of the vibration signal. If this indicator abnormally increases, it indicates that there are hidden dangers in the processing process. The margin indicator is the ratio of vibration peak to square root amplitude, reflecting the impact characteristics of the vibration signal. The larger the indicator, the stronger the processing vibration impact. By comparing the evaluation indicators, it can be seen that double-sided collaborative machining has greatly improved the vibration compared to unsupported single-sided machining, indicating that the double-sided collaborative machining method can improve the machining result effectively.

Table 4. Evaluation indicators data of deformation and vibration evaluation.

Machining Method	Deformation Range (mm)	X_p (mm)	PP (mm)	rms (mm)	CL
Single-sided	−0.209~0.456	0.0712	0.1430	0.0081	19.7095
Double-sided Collaborative	−0.193~0.138	0.0441	0.0871	0.0034	16.0946

Finally, the quality of the machined blade is evaluated in the experiment by measuring the blade radius, section length under three radii, and surface roughness. Measure the blade radius length R and the section length at 0.2R, 0.5R, and 0.95R using the range finder, and take the average of three times as the final result, as shown in Table 5. The deviation rate of blade radius is 0.13%, which meets the requirement of ±0.2% for extreme radius tolerance of Class S in ISO 484-2:2015 [19]. The deviation rates of section length are 0.38%, 0.46%, and 0.35%, respectively, meeting the requirement of +2% for blade section length tolerance of Class S in ISO 484-2:2015 [19].

Table 5. Measurement data of blade profile dimensions.

Measurement Target	Ideal Size (mm)	Measurement 1	Measurement 2	Measurement 3	Average Size (mm)	Deviation Value (mm)
Blade radius	229.36	229.69	229.62	229.64	229.65	0.29
Section length at 0.2R	106.89	107.32	107.28	107.29	107.30	0.41
Section length at 0.5R	142.18	142.80	142.83	142.88	142.84	0.66
Section length at 0.95R	17.17	17.22	17.28	17.20	17.23	0.06

The surface roughness of the area at 0.4R of the blade is measured by laser rangefinder. The evaluation length is composed of five sampling lengths of 0.8 mm, and the roughness curves shown in Figure 12 are obtained by processing the contour curve. The blue color is the roughness curve, and the red color is the arithmetic mean center line. Calculate the arithmetic mean deviation value Ra of the sampling length by Equation (26):

$$\mathrm{Ra} = \frac{1}{m}\sum_{i=1}^{m}|h_m|, \tag{26}$$

where m is the number of measuring points within the sampling length, and h_m is the fluctuation value of the distance from a single measuring point to the arithmetic mean center line.

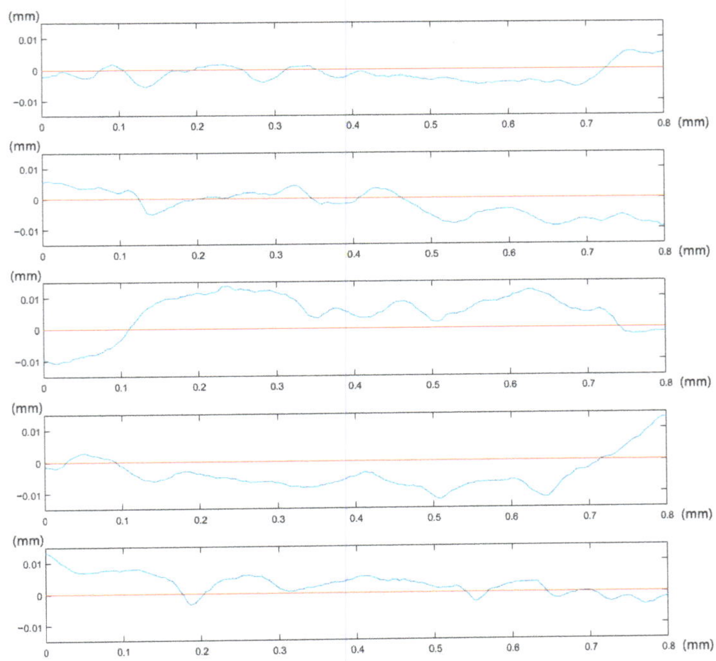

Figure 12. Blade roughness curve.

The arithmetic mean deviation Ra of the five sampling lengths is 0.0026 mm, 0.0041 mm, 0.0072 mm, 0.0058 mm, and 0.0037 mm, respectively, and the arithmetic mean deviation Ra of the evaluated length is 0.00468 mm, meeting the requirements of 5 μm for surface roughness arithmetic mean deviation of Class II in ISO 484-2: 2015.

8. Conclusions

In response to the problems of thin edges and complex shapes of propeller blades, the surface of the propeller blade is divided into four regions and adopts targeted machining methods to achieve double-sided collaborative machining, while avoiding the problem of collision between cutters on both sides during the machining process. However, the machining of the tip region still requires auxiliary shoring to decrease deformation and vibration, which urgently needs to be solved.

1. The spatial curve of the tool path is expressed in the form of parameter equations by fitting the projected curve of the tool path on the Y-X and Y-Z planes. The cutter position points are interpolated on the tool paths of the template surface through double-curve interpolation, and the data alignment from the template surface to the projection surface is completed by calculating the length ratio of the cutter position points in the tool path. This method can quickly obtain the symmetrical-cutter-position-points while ensuring uniformity and smoothness.
2. The interpolation of cutter position points in the variable region was completed by the sigmoid deformation curve, achieving the conversion between double-sided symmetrical machining and double-sided staggered machining. The variable speed stage during the machining process is stable and reliable through experimental verification, suitable for frequent variable speed operations.
3. The feasibility and superiority of the double-sided collaborative machining tool path planning algorithm proposed in this paper for propeller blades have been verified by practical machining experiments. Comparing the integrated data results of double-sided collaborative machining with single-sided machining, the former greatly improves the deformation and vibration in propeller blade machining. By measuring and evaluating the quality of the blades, it can be concluded that it meets the general accuracy requirements, proving that the double-sided collaborative machining method has broad applied foreground and practical value.

Author Contributions: Conceptualization, R.W.; methodology, R.W. and Y.G.; software, X.G. and Y.G.; validation, Y.G.; formal analysis, R.W.; investigation, R.W.; resources, R.W.; data curation, Y.G.; writing—original draft preparation, Y.G.; writing—review and editing, R.W. and X.G.; visualization, Y.G.; supervision, R.W.; project administration, R.W.; funding acquisition, R.W. All authors have read and agreed to the published version of the manuscript.

Funding: This work was supported by the National Natural Science Foundation of China (51975157): Research on the improvement of machining precision on double-cutter & double-surface symmetrical machining for large scale marine propeller.

Institutional Review Board Statement: Not applicable.

Informed Consent Statement: Not applicable.

Data Availability Statement: Data is unavailable due to privacy.

Conflicts of Interest: The authors declare no conflict of interest.

References

1. Gui, L.; Li, S. Status and Development Trend of Domestic Heavy-duty CNC Machine Tools. *China Mech. Eng.* **2020**, *31*, 2780. [CrossRef]
2. Yan, M. The World's Largest Propeller Processing Equipment Has Emerged. *China Ship News*, 13 July 2007; p. 007.
3. Mejbel, M.K.; Abdullah, I.T.; Taieh, N.K. Thin Wall Manufacturing Improvement using Novel Simultaneous Double-Sided Cutter Milling Technique. *Int. J. Automot. Mech. Eng.* **2022**, *19*, 9519–9529. [CrossRef]
4. Fu, R.; Curley, P.; Higgins, C.; Kilic, Z.M.; Sun, D.; Murphy, A.; Jin, Y. Double-sided milling of thin-walled parts by dual collaborative parallel kinematic machines. *J. Mater. Process. Technol.* **2022**, *299*, 117395. [CrossRef]
5. Mori, T.; Hiramatsu, T.; Shamoto, E. Simultaneous double-sided milling of flexible plates with high accuracy and high efficiency—Suppression of forced chatter vibration with synchronized single-tooth cutters. *Precis. Eng.* **2011**, *35*, 416–423. [CrossRef]
6. Shamoto, E.; Mori, T.; Nishimura, K.; Hiramatsu, T.; Kurata, Y. Suppression of regenerative chatter vibration in simultaneous double-sided milling of flexible plates by speed difference. *CIRP Ann.* **2010**, *59*, 387–390. [CrossRef]
7. Shamoto, E.; Mori, T.; Sencer, B.; Suzuki, N.; Hino, R. Suppression of regenerative chatter vibration in multiple milling utilizing speed difference method–Analysis of double-sided milling and its generalization to multiple milling operations. *Precis. Eng.* **2013**, *37*, 580–589. [CrossRef]

8. Hao, J.; Zhao, Y.; Wang, H.; Sheng, D. Synthetical Stiffness Analysis and Optimization of Mirror Support Mechanism for Thin-walled Structures. *Mach Des. Res.* **2015**, *31*, 155–159. [CrossRef]
9. Song, D.; Xue, F.; Zhang, J.; Zou, C.; Zhao, W.; Lu, B. Tool Path Smoothing Algorithm for Nine-Axis Synchronized Twin-Tool Milling of Blades. *J. Xi'An Jiaotong Univ.* **2019**, *53*, 1745–1757. [CrossRef]
10. Lu, B.; Fang, Y.; Xu, D.K.; Chen, J.; Ai, S.; Long, H.; Ou, H.; Cao, J. Investigation of material deformation mechanism in double side incremental sheet forming. *Int. J. Mach. Tools Manuf.* **2015**, *93*, 37–48. [CrossRef]
11. Bo, Q.L. Research on Stability Monitoring of Large Thin-walled Parts Mirror Milling. Ph.D. Thesis, Dalian University of Technology, Dalian, China, November 2019.
12. Mandaloi, G.; Nagargoje, A.R.; Mall, A.P.; Gupta, A.K.; Tiwari, M.K.; Dubey, A.K.; Tandon, P. Feature-Based Double-Sided Deformation Machining Approach for Manufacturing Freeform Monolithic Components. *Int. J. Adv. Manuf. Technol.* **2022**, *123*, 2107–2121. [CrossRef]
13. Zhang, Q.B. Research on Double-Cutter and Double-Surface Machining Technology based on XYZ-3RPS Serial-Parallel Mechanism. Master's Thesis, Harbin Institute of Technology, Harbin, China, June 2015.
14. Wang, R.; Zhang, Q.B. Post-Processing Algorithm Oriented to XYZ-3RPS Serial-Parallel Machine Tool. *J. Harbin Inst. Technol.* **2016**, *48*, 130–134. [CrossRef]
15. Budak, E.; Comak, A.; Ozturk, E. Stability and high performance machining conditions in simultaneous milling. *CIRP Ann.* **2013**, *62*, 403–406. [CrossRef]
16. Wu, P.C. CFD Body Force Propeller Model with Blade Rotational Effect. *Appl. Sci.* **2022**, *12*, 11273. [CrossRef]
17. Jiang, X.Y. Tool Path Planning of Moving along the Fluidflow Direction on Propeller Blade Surface in Machining Process. Ph.D. Thesis, Huazhong University of Science and Technology, Wuhan, China, January 2014.
18. Chen, L.J.; Sui, Y.Z.; Wang, Z.Z.; Wei, G.X. Planning and Controlling Method of Feedrate in Five-axis Linear Interpolation. *J. Beijing Univ. Technol.* **2020**, *62*, 979–987+1017. [CrossRef]
19. *ISO 484-2:2015*; Shipbuilding–Ship Screw Propellers–Manufacturing Tolerances–Part 2: Propellers of Diameter between 0.80 and 2.50 Minclusive. International Standards Organization: Geneva, Switzerland, 2015.

applied sciences

MDPI

Article

The Variation of Surface Shape in the Gas Jet Forming

Xinming Zhang [1,2,3], Mingwei Wang [2,3] and Weijie Fu [1,2,3,*]

1 School of Mechatronic Engineering and Automation, Foshan University, Foshan 528001, China
2 Ministry of Education Key Laboratory for Cross-Scale Micro and Nano Manufacturing, Changchun 130022, China
3 School of Mechatronic Engineering, Changchun University of Science and Technology, Changchun 130022, China
* Correspondence: weijie.fu@hotmail.com

Abstract: This study investigated the the gas jet forming process for optical aspherical mirror blanks. The trend of the influence of the gas jet parameters on the surface shape of the mirror blanks was inferred by analysing the variation in the morphology of the gas jet stream. Based on the theoretical analysis, the inference was verified by numerical simulation. The experimental results were compared with the simulation predictions, and it was found that the average prediction deviation for the diameter was 1.07 mm, while the average prediction deviation for the principal curvature was 0.03665 mm^{-1}, which is challenging to correct in simulation. Therefore, we developed a dimensionless prediction model of the surface curvature and surface diameter of the mirror blanks by considering the jet parameters using experimental data. The model's average prediction error for the surface diameter of the formed surface was 0.3192 mm, and the average prediction error of the principal curvature for the formed surface was 0.00269 mm^{-1}.

Keywords: gas jet forming; mirror blank; prediction models for surface shape; optical processing

check for updates

Citation: Zhang, X.; Wang, M.; Fu, W. The Variation of Surface Shape in the Gas Jet Forming. *Appl. Sci.* **2023**, *13*, 504. https://doi.org/10.3390/app13010504

Academic Editors: Ivan Galvão, Rui Manuel Leal and Gustavo H. S. F. L. Carvalho

Received: 4 December 2022
Revised: 27 December 2022
Accepted: 27 December 2022
Published: 30 December 2022

1. Introduction

With the development of technology, aspherical mirrors are widely used in aerospace, medical testing, and information communication, and are among the core components of modern optical systems. For instance, the Giant Magellan Telescope developed by the University of Arizona and the telescopes developed by NASA all employ high sub-aspherical mirrors [1,2]. Mirror blanks are the substrates used for optical mirror coatings, and their surface accuracy and quality determine the optical performance of the mirror after coating. Therefore, processing mirror blanks is the most critical aspect of optical mirror formation [3,4].

The conventional aspherical mirror processing methods are material removal formation [5], compression moulding [6], and spin-casting [7]. These methods generally have high processing costs, low flexibility, and limited surface shape. Fu et al. proposed a the gas jet forming method for processing mirror blanks [8]. Under this air pressure, the pool of photosensitive resin is deformed by the impact and cured when the desired surface shape is achieved. This method allows the desired optical surface to be obtained by controlling the nozzle height and the gas flow rate.

Currently, the studies related to the gas jet forming in the field of mirror blank processing production are still in their infancy. However, the the gas jet forming process has become one of the hot topics in the field of multiphase fluids. In 1965, Banks et al. observed the liquid surface dimple phenomenon caused by cylindrical jets [9]. In 1970, Molloy et al. confirmed that the profile of liquid surface dimples is related to the jet velocity and nozzle diameter [10]. In 2004, Meidani et al. used an experimental setup of liquid metal systems to study the influence of gas flow rate and nozzle height on the depth of liquid surface dimples [11]. In 2012, Ek et al. experimented with impacting liquid steel with a gas jet and established a depth prediction model related to the gas flow rate and nozzle height [12]. In 2017, Kalifa et al. used the Reynolds stress model (RSM) to study the interaction process of a gas jet perpendicular to a liquid interface [13]. In the same year, Jurman et al. derived a nonlinear wave equation capable of quantitatively describing the surface waves caused by airflow [14]. According to the above studies, the gas jet forming process mainly consider the interaction mechanism of the gas–liquid interface, the surface wave phenomenon, and the depth of the liquid surface dimples. These studies cannot predict the diameter and principal curvature of the mirror blank. Therefore, it is necessary to establish the relationship between the gas flow rate, nozzle height, and principal curvature and surface diameter of the mirror blank for further study of the jet formation process.

According to the research completed so far, this paper investigates the change rule of the principal curvature and surface diameter of mirror blanks in the gas jet forming. Firstly, the influence of the gas jet parameters on the pressure distribution rule in the cross-section of the gas stream is analysed, and a model of the pressure distribution in the initial section of the gas stream is established. Based on this gas pressure distribution model, an inference is drawn from the relationship between the gas jet parameters and the surface parameters. A numerical simulation is carried out to understand this inference. Finally, the inference is verified by the gas jet forming experiments. The dimensionless mathematical model of the principal curvature and surface diameter of the mirror blank with respect to the jet parameters is established by analysing the causes of the calculation deviations.

2. Principles of the Gas Jet Forming and Related Inferences

2.1. Principles of the Gas Jet Forming

In the the gas jet forming process, the jet from the nozzle, due to the entrainment effect, increases the end surface of the stream. At the same time, the flow velocity of the jet also decreases, and the deformed gas jet carrying kinetic energy impacts the surface of the liquid pool. The velocity distribution of the jet in the cross-section is characterised by a maximum velocity at the centre line of the flow stream, and the further away from the centre axis in the same cross-section, the lower the gas flow velocity. In this case, the liquid pool's impingement position cannot bear the impact pressure generated by the gas jet, and the pressure difference between the inside and the outside will finally form a stable surface. According to mechanical analysis, the surface shape formed by the gas jet profile is determined by the air pressure, hydrostatic pressure, and surface tension generated by the jet, and the equilibrium composed of the above forces is shown in Figure 1 [15], where p is the interfacial pressure difference, τ is the interfacial stress caused by the jet, and dh and dr are the ds interfacial unit components.

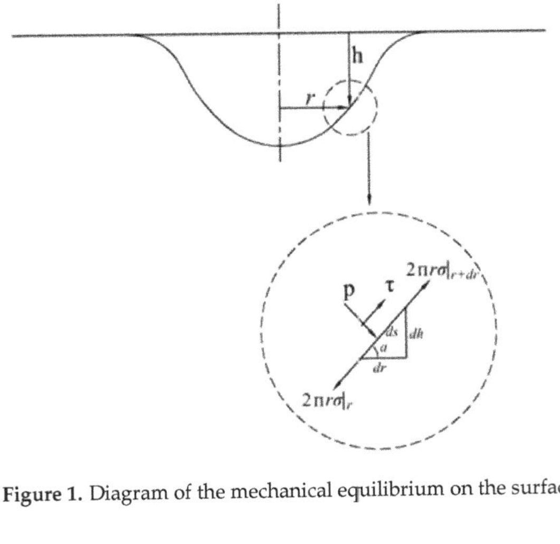

Figure 1. Diagram of the mechanical equilibrium on the surface formed by the gas jet.

2.2. Inference of the Relationship between Jet Parameters and Mirror Blank Parameters

According to the equilibrium theory described in Section 2.1, the surface profile of the mirror blank is closely related to the interfacial pressure and the physical properties of the photosensitive resin. The physical properties of the photosensitive resin used in this study are stable, and by changing the velocity of the gas jet and the nozzle's height from the liquid surface, the pressure on the surface of the resin pool can be altered to achieve effective control of the shape of the resulting surface. Therefore, it is important to analyse the gas flow field created by the gas jet to determine the pressure distribution on the surface of the liquid photosensitive resin.

According to previous studies [16], the velocity distribution of the gas flow field on the axis of the flow stream can be divided into a potential core region and a self-similarity region, with the BOE as the transition zone, as shown in Figure 2.

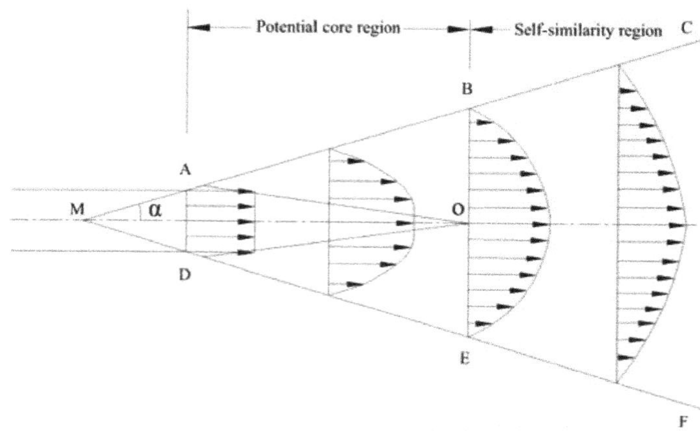

Figure 2. Schematic diagram of the changes in the gas flow stream.

The distance of the potential core region S_n is determined by Equation (1) [17].

$$S_n = 0.671\frac{r_0}{\alpha}, \tag{1}$$

where r_0 is the radius of the nozzle and α is the diffusion angle; in this paper, α is 0.08.

As this paper is concerned with the effects of the jet velocity and the nozzle height from the liquid surface on the mirror profile, the nozzle diameter was taken as a fixed quantity. In this paper, the nozzle diameter chosen was 10 mm, and the above parameters were substituted into Equation (1) to obtain the following calculation result:

$$S_n = 0.671\frac{r_0}{\alpha} = 0.671 \times \frac{0.005}{0.08} = 41.93\ \text{mm}, \tag{2}$$

According to the analysis results shown in Figure 2, the gas flow field is divided by the BOE turning surface, and there is a variation trend between the potential core region and the self-similarity region. According to the calculation results obtained from Equation (2), when the self-similarity region is selected for the gas jet forming, the nozzle's height from the liquid surface is greater than 41.93 mm. At this point, it can be found through the experiment (as shown in Figure 3) that the surface formed is very shallow, and it is not easy to observe the influence of the changes in the gas jet parameters on the surface shape of the mirror blank.

Figure 3. Experimental results of the gas jet forming using the self-similarity region.

Therefore, the potential core region was chosen as the formation flow field in this study. According to Figure 2, the velocity of the flow field of the potential core region in the cross-section is the same after the gas is ejected from the nozzle, and the diameter of the core region's cross-section gradually decreases with the movement of the gas stream. In 1962, Banks et al. described the main section of the jet flow field with the following equation [18]:

$$P = \frac{1}{2}\rho_G V_0^2\left(\frac{d}{h}\right)\cdot e^{-c(\frac{r}{h})^2}, \tag{3}$$

where P is the pressure generated by the gas flow, ρ_G is the gas density, V_0 is the gas flow rate, d is the nozzle diameter, h is the nozzle height, and c is a constant.

However, the above equation can only describe the flow field distribution in the self-similarity region, but not the gas flow field in the potential core region. Comparing the self-similarity region with the flow field distribution law of the potential core region, we can see that there is a jet-preserving core region in the starting section, and there is a difference in the number of times describing the normal distribution e. Therefore, it is important to develop a mechanistic model suitable for describing the flow field distribution in the potential core region and analysing the flow field pressure variations. In this study, a mechanistic model describing the flow field in the potential core region was established and modelled as follows:

$$\begin{cases} P = \frac{1}{2}\rho_G V_0^2 & r < r_{cor} \quad (a) \\ P = \frac{1}{2}\rho_G V_0^2\left(\frac{d}{h}\right)\cdot e^{-c(\frac{rh}{\sigma})^2} & r > r_{cor} \quad (b) \end{cases} \tag{4}$$

where r_{cor} is the core zone radius, $r_{cor} \propto (1/h)$, and σ is a constant (mm^2). Equation (4) (a) describes the pressure distribution within the core zone, while Equation (4) (b) describes the pressure distribution outside the core zone to the boundary layer.

According to Section 2.1 of this paper, the surface shape of the mirror blank is formed by the combined action of air pressure, surface tension, and the hydrostatic pressure of the liquid. Of these, the air pressure has a positive effect on the expansion of the surface. In contrast, the surface tension and static pressure have a negative effect by inhibiting the expansion of the surface. During the the gas jet forming process, the air pressure is stronger than the combined inhibitory effect of the surface tension and static pressure, causing a tendency for the surface profile to expand outwards until the three forms of action reach equilibrium. For this reason, Equation (4) can be modified to obtain Equation (5).

$$P = \frac{\rho {V_0}^2 d}{2he^{c(\frac{rh}{\sigma})^2}}, \tag{5}$$

According to Equation (5), the magnitude of the pressure acting on the surface of the liquid cell is related to the initial jet velocity V_0. The greater the V_0, the greater the resulting pressure P, resulting in a larger diameter of the surface formed by the gas jet. It can be inferred that the diameter formed by the gas jet increases as the initial gas flow rate V_0 increases. In addition to velocity, the pressure is also related to the nozzle's height from the liquid surface. Although the cross-sectional diameter of the flow stream increases as the nozzle's height from the liquid surface increases, the greater the nozzle height, the lower the pressure within the radial cross-section, resulting in a smaller diameter formed by the gas jet.

The principal curvature is one of the key parameters describing the profile of an optical mirror, and it is not correlated with the diameter. Therefore, it is also necessary to analyse the law of change in the principal curvature of the surfaces formed in the gas jet forming. Due to the volume reduction in the potential core region and the effect of the hydrostatic pressure of the liquid, the principal curvature is determined by both the pressure in the potential core region and the pressure distribution generated by the flow streams in the vicinity of the core.

By taking the derivative of P with respect to r in Equation (4) (b):

$$\frac{dP}{dr} = -cr\sigma^{-2}\rho_G hdV_0^2 \cdot e^{-c\left(\frac{rh}{\sigma}\right)^2}, \tag{6}$$

The second derivative of the above equation:

$$\frac{d^2P}{dr^2} = c\sigma^{-2}\rho_G V_0^2 dh \cdot e^{-c\left(\frac{rh}{\sigma}\right)^2}\left(2ch^2\sigma^{-2}r - 1\right), \tag{7}$$

According to the mathematical equation, the curvature K at each point of the airflow can be obtained from Equation (8).

$$K = \frac{\left|\frac{d^2P}{dr^2}\right|}{\left[1 + \left(\frac{dP}{dr}\right)^2\right]^{\frac{3}{2}}}, \tag{8}$$

Substituting Equations (6) and (7) into Equation (8) yields the curvature equation describing the pressure distribution of the gas jet acting on the fluid surface.

$$K = \frac{\left|c\sigma^{-2}\rho_G V_0^2 dh \cdot e^{-c\left(\frac{rh}{\sigma}\right)^2}\left(2ch^2\sigma^{-2}r - 1\right)\right|}{\left[1 + \left(-cr\sigma^{-2}\rho_G hdV_0^2 \cdot e^{-c\left(\frac{rh}{\sigma}\right)^2}\right)^2\right]^{\frac{3}{2}}}, \tag{9}$$

The focus of this paper is on the preserving core region and the pressure distribution near the preserving core region ($r \to 0$). At this point, the principal curvature equation of the airflow stream is shown in Equation (10).

$$K = \lim_{r\to 0} K = \frac{\left|c\sigma^{-2}\rho_G V_0^2 dh \cdot e^{-c\left(\frac{rh}{\sigma}\right)^2}\left(2ch^2\sigma^{-2}r - 1\right)\right|}{\left[1 + \left(-cr\sigma^{-2}\rho_G hdV_0^2 \cdot e^{-c\left(\frac{rh}{\sigma}\right)^2}\right)^2\right]^{\frac{3}{2}}} = c\sigma^{-2}\rho_G V_0^2 dh, \tag{10}$$

Analysis of Equation (10) can lead to the following inference: when the nozzle diameter is determined, the principal curvature and the initial velocity of the gas stream change with the nozzle's height from the liquid surface. In contrast, the principal curvature of the mirror blank can be inferred from the increasing initial velocity of the jet to increase with the increasing height of the nozzle from the liquid surface.

3. Numerical Simulation

Although the relationship between the jet parameters and the mirror blank parameters is inferred above, it is only mechanistic. In order to verify the above inference, the relationship between the gas jet parameters and the surface shape parameters of the formed surface was analysed using fluent software to investigate the variation of the diameter and the principal curvature of the formed surface due to the variation of the gas flow and nozzle height.

3.1. Establishment of the Model

The simulation model of the gas injection moulding process is shown in Figure 4, where "inlet" is the gas inlet and "wall" is the wall of the nozzle and the wall of the mould. h_0 is the height of the glass dish wall, and h_1 is the liquid phase's liquid level. The simulation was calculated using a two-dimensional axisymmetric computational domain with the axis of symmetry shown in the figure.

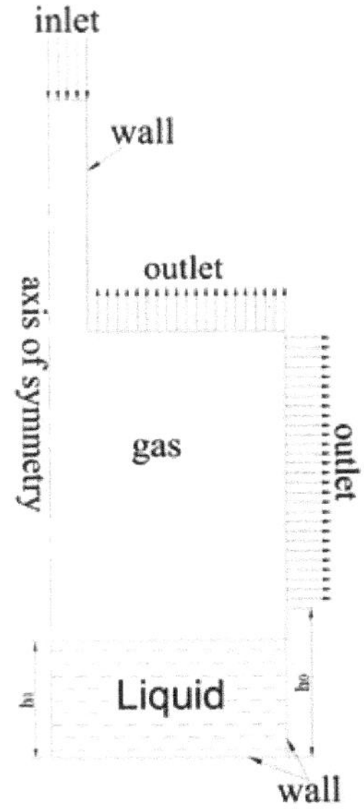

Figure 4. The model diagram of the gas jet forming.

3.2. Control Equations and Discrete Methods

The VOF interface tracking model in the Reynolds-averaged Navier–Stokes framework was used in the simulation to predict the the gas jet forming process. Nichols and Hirt first proposed the VOF method in 1975 to capture free surface calculations through a computational grid [19]. The fluid model was refined by these two researchers in 1981 and is still in use today [20].

The numerical simulation of the the gas jet forming method is a multinomial fluid model. During the solution process, the gas and liquid phases fill the calculation domain, and in this model both the gas and liquid phases are continuous, satisfying the Navier–Stokes equations. Due to the low influence of temperature on the experiment, the energy conservation factor is ignored in the calculations. The mass equation is shown in Equation (11), and the momentum equation is shown in Equation (12) [21].

$$\frac{\partial \rho}{\partial t} + \nabla \cdot (\rho U) = 0, \tag{11}$$

$$\frac{\partial \rho U}{\partial t} + \nabla \cdot (\rho U U) - \nabla \cdot \tau = -\nabla p + \rho g + F, \tag{12}$$

where t is the time, U is the gas flow rate, p is the pressure, g is the gravity vector, μ is the viscosity, ρ is the density, and F is the source term, which is provided in the experiments of the jet spray moulding method, including surface tension.

According to Section 2.1 above, surface tension is an important influencing factor for the prediction of the mirror blank's surface shape. The surface tension model in this paper is the continuous surface force model proposed by Brackbill. As shown in Equation (13), in this model, the surface tension is added to the source term of the momentum equation [22].

$$F_{CSS} = \nabla \cdot T, \tag{13}$$

Expressing the stress surface tensor T [22]:

$$T = \sigma(|\nabla\alpha|I - \frac{\nabla\alpha \otimes \nabla\alpha}{|\nabla\alpha|}), \tag{14}$$

where I is the stress tensor, α is the volume fraction, σ is the surface tension coefficient, $\otimes$ is the tensor product of the two vectors, and n is the volume fraction gradient.

Within the grid of the VOF method, the two phases share a common set of momentum equations. Therefore, the rate properties and the material properties within the grid are jointly determined by the physical properties of the two phases, along with the volume fraction [23].

The formula for the mixing density ρ_m is as follows [23]:

$$\rho_m = \alpha_c\rho_c + \alpha_d\rho_d, \tag{15}$$

where α_c is the volume fraction of the gas phase, ρ_c is the physical density of the gas phase, α_d is the volume fraction of the liquid phase, and ρ_d is the physical density of the liquid phase.

The formula for the mixing viscosity μ_m is as follows [23]:

$$\mu_m = \alpha_c\mu_c + \alpha_d\mu_d, \tag{16}$$

where μ_c is the physical viscosity of the gas phase, $\mu_{d'}$ is the physical viscosity of the liquid phase. The grid rate $\mathbf{U_m}$ property has the following formula [23]:

$$\mathbf{U_m} = \frac{\alpha_c\rho_c\mathbf{U}_c + \alpha_d\rho_d\mathbf{U}_d}{\rho_m}, \tag{17}$$

where $\mathbf{U_C}$ is the gas phase rate property at the last time step of the grid, while $\mathbf{U_d}$ is the liquid phase rate property at the last time step of the grid.

In order to ensure the better stability of the VOF, an explicit algorithm is used in the time domain, the transient solution is used for the numerical simulation process, and the PISO algorithm is applied to solve the pressure-rate field. The first-order upwind algorithm is used to discretise the momentum equation, the PRESTO algorithm is used to discretise the pressure space, and the least squares cell-based algorithm is used to discretise the gradient space.

3.3. Grid Independence Test

The simulation's mesh quality directly determines the numerical simulation's prediction accuracy. Therefore, a grid independence test is required before calculation. This paper uses a rectangular orthogonal grid to partition the model. In the the gas jet forming process, two cases of surface profiles are calculated at different grid densities, namely, no-point and two-point, as shown in Figure 5.

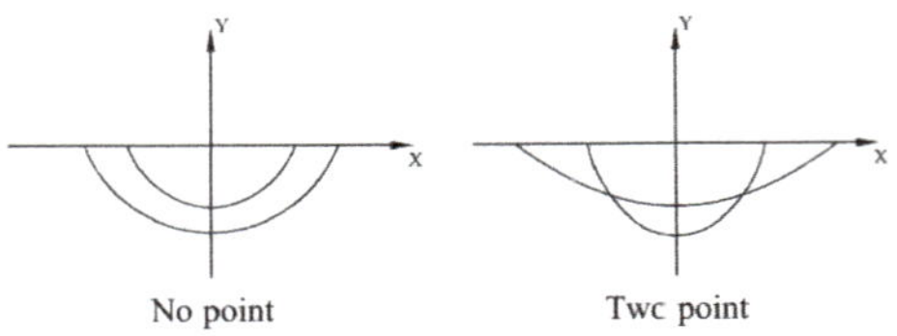

Figure 5. Prediction of curves occurring at different grid densities.

There are many methods fcr grid independence tests, such as Harsdorf's method [24], Freshet's method [25], and LCSS [26]. However, these methods are only suitable for judging straight-line trajectories and cannot achieve the expected results in the case of judging similar curves. For this reason, this paper proposes a new method for determining the similarity of specular contours for the surface profile characteristics formed by the gas jet forming. Its similarity formula is as follows:

$$\delta = \frac{C}{A + B - C}, \tag{18}$$

where A is the area of intersection of curve 1 with the X axis, B is the area of intersection of curve 2 with the X-axis, and C is the common area of A and B.

The grid size was set to 0.7 mm, 0.6 mm, 0.5 mm, 0.4 mm, 0.3 mm, 0.2 mm, 0.1 mm, and 0.09 mm to mesh the model. The number of meshes corresponding to the above mesh sizes was 3565, 4850, 6917, 10,911, 19,110, 42,596, 169,091, and 209,038, respectively. A curve similarity equation (Equation (18)) was applied to calculate the similarity of two adjacent curves, and the point line diagram was drawn based on the calculation results, as shown in Figure 6.

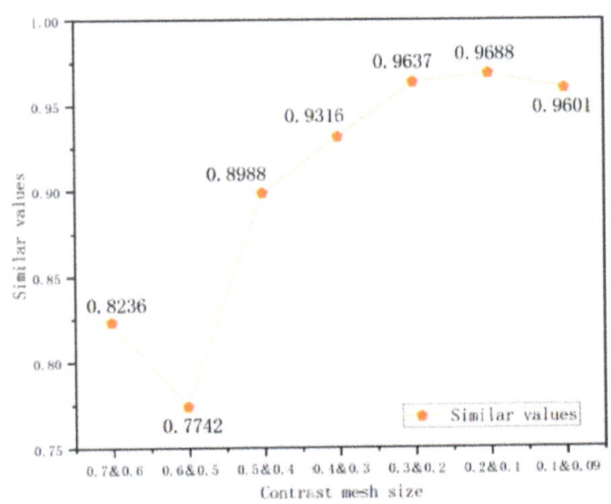

Figure 6. The results of the grid independence test.

According to the results of the calculations, the similarity values fluctuate considerably when the grid (or grids) size is between 0.7 mm and 0.3 mm. Between 0.3 mm and 0.09 mm, the fluctuations in the curve similarity values became smooth and reached over 95%. The similarity value did not reach 100%, due to a slight error caused by the time step adjustment. Based on the grid independence experiments, 0.3 mm was chosen as the grid size for the numerical simulations.

3.4. The Setting of the Numerical Simulation Parameters and Analysis of the Results

According to Section 2.2, it can be seen that when the nozzle is too high from the liquid surface, it will make the formation difficult to observe macroscopically. In contrast, a low nozzle distance from the liquid surface will cause waves at the gas–liquid interface. Therefore, the nozzle height was set between 20 mm and 40 mm, and the gas jet velocity was between 9.5 m/s and 11 m/s. The jet parameters for the numerical simulation and the subsequent experiments were used in 4 × 5 orthogonal experiments, and the gas jet parameters were set as shown in Table 1.

Table 1. Gas jet parameters.

Parameters	Numerical Value
Gas flow rate (m/s)	9.5, 10, 10.5, 11
Nozzle height (mm)	20, 22.5, 25, 27.5, 31

The material used to form the mirror blanks was photosensitive resin, and the gas emitted from the nozzle was air. The relevant properties of the photosensitive resin were tested using a DA-130N densitometer, a VT-06 viscometer, and a SITA surface tension meter, and the results are shown in Table 2.

Table 2. Material properties.

<table>
<tr><th></th><th>Air</th><th>Photosensitive Resin</th></tr>
<tr><td>Density/kg·m^{-3}</td><td>1.225</td><td>977</td></tr>
<tr><td>Viscosity/cps</td><td>1.7894×10^{-5}</td><td>2000</td></tr>
<tr><td>Surface tension/mN·m^{-1}</td><td colspan="2">37.6</td></tr>
</table>

The final mirror blank profile curves for the different gas jet parameters were obtained by numerical simulation, as shown in Figure 7, where the yellow part is the air and the red part is the photosensitive resin liquid pool. By collecting the calculated data from the numerical simulation, the relationship between the gas flow rate and the diameter of the forming surface was determined, as shown in Figure 8a, and the relationship between the distance from the nozzle to the liquid surface and the diameter of the formed surface is shown in Figure 8b.

Figure 7. Numerical simulation results of the the gas jet forming.

As shown in Figure 8, the diameter of the surface formed by the gas jet increased as the gas flow rate increased when the nozzle's height from the liquid surface was fixed. When the gas flow rate was fixed, the diameter of the surface formed by the gas flow rate decreased as the nozzle height increased. When the gas flow rate was between 9.5 m/s and 11 m/s and the nozzle's height from the liquid surface was between 20 mm and 30 mm, the minimum diameter of the formed surface was 18.6904 mm and the maximum was 30.8936 mm in the simulations.

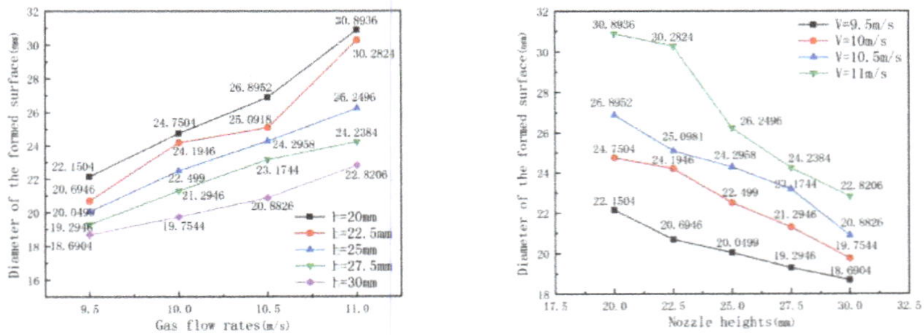

(a) (b)

Figure 8. The diameter of the formed surface with different gas flow rates and nozzle heights in simulations. (**a**) is the relationship between the gas flow rate and the diameter of the forming surface. (**b**) is the relationship between the nozzle height and the diameter of the formed surface.

The principal curvature of the mirror blank is challenging to determine directly from the profile information obtained from the numerical simulation, unlike the diameter of the surface formed by the gas jet. Therefore, using the surface profile data calculated from the simulation, the principal curvature was calculated using a data-fitting method. The shape equation for fitting the data is given in Equation (19) [27].

$$z = \frac{Cr^2}{1 + \sqrt{1 - (K+1)C^2r^2}} + ar^4 + br^6 + \cdots - h_m, \quad (19)$$

where C is the principal curvature of the mirror blanks, K is the cone coefficient, $a, b \ldots$ is a multinomial coefficient, and h_m is the lowest point of the mirror blank. The principal curvature of the formed surface, simulated for different gas flow rates and nozzle heights from the liquid surface, is shown in Figure 9.

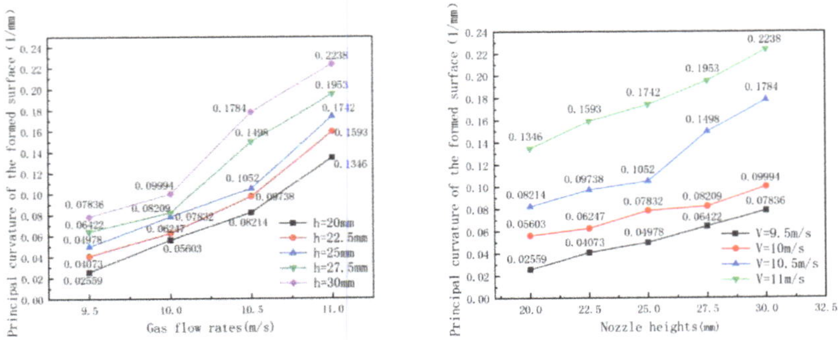

(a) (b)

Figure 9. The principal curvature of the formed surface with different gas flow rates and nozzle heights in simulations. (**a**) is the relationship between the gas flow rate and the principal curvature of the formed surface. (**b**) is the relationship between the nozzle height and the principal curvature of the formed surface.

As shown in Figure 9, both the gas flow rate and the nozzle height from the liquid surface play a significant role in the changes in the principal curvature of the mirror blank. As the nozzle height increases, the principal curvature increases. As the gas flow rate increases, the principal curvature increases. Within the range of jet parameters set in this paper, when the jet velocity was 11 m/s and the nozzle's height from the liquid surface was 30 mm, the principal curvature of the formed surface reached its maximum, at 0.1159 mm^{-1}. When the gas flow rate was 9 m/s and the distance of the nozzle from the liquid surface was 20 mm, the principal curvature of the formed surface was minimal, at 0.03723 mm^{-1}.

4. Experiments

The relevant experiments were carried out by building a the gas jet forming testing platform to verify the effectiveness of the above inferences and simulation results. The details are as follows:

The Gas Jet Forming Method

The gas jet forming of the mirror blanks was accomplished by combining the formation of the desired surface shape by the gas jet and the UV-catalysed curing of the photosensitive resin. The forming process was divided into four stages, as shown in Figure 10.

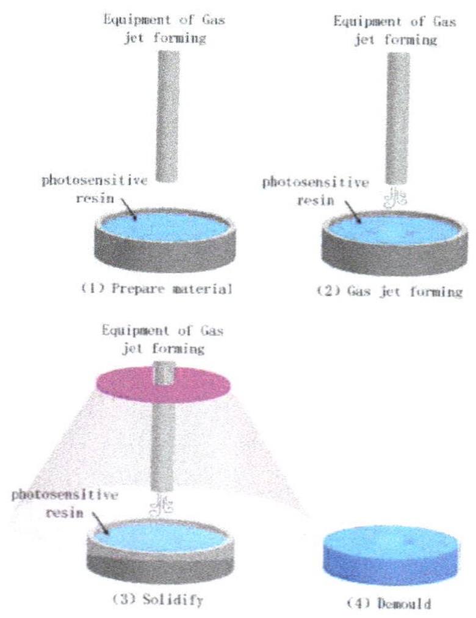

Figure 10. Schematic diagram of the process of the gas jet forming.

The experimental setup used in this research is shown in Figure 11. The experimental setup was adjusted for the nozzle axis position and the vertical height of the nozzle mouth from the epoxy resin surface through a precision three-axis moving platform with an accuracy of 0.01 mm.

During the experiment, the air was injected into the coupled gas tank (2) via the air compressor (1). The compressed air was released from the tank and filtered through the oil–water separator (3) and the refrigeration dryer (4) to ensure that the gas used in the experiments was clean and dry. The gas pressure was regulated via the gas MFC (5) to ensure a stable air pressure of the gas supplied during the experiment. The flow rate of the gas jet through the nozzle during the experiment was controlled using a gas jet controller (6). The experimental gas was finally jetted through the precision electroplated nozzle (8) onto the surface of the liquid epoxy resin in a transparent glass mould (10). The transparent glass mould containing the liquid epoxy resin was clamped to a marble platform (11), and the spatial position of the nozzle during the experiment was controlled by a precision three-axis moving platform (9). The forming process of the optical surface was directly observed through the transparent glass mould and recorded with a high-speed camera. The mirror blanks formed by solidifying the epoxy resin were measured using a surface profiler (Taylor Hobson PGI) on the resulting surface.

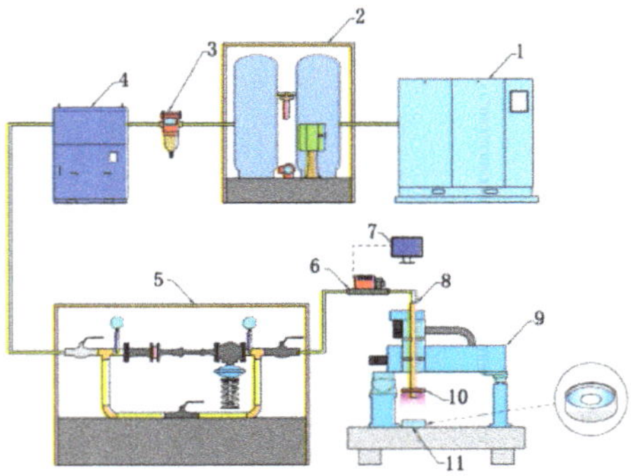

Figure 11. Setup of the experimental the gas jet forming platform: (1) air compressor; (2) coupled gas tank; (3) water–oil separator; (4) refrigeration dryer; (5) gas MFC; (6) gas jet controller; (7) upper computer; (8) precision electroplated nozzle; (9) precision three-axis moving platform; (10) UV light source; (11) marble platform.

5. Experimental Results

In this study, we used the equipment described above to experiment with the the gas jet forming of mirror blanks. In order to compare the experimental results with the numerical simulation results, the the gas jet forming parameters were kept consistent with the numerical simulations. The gas flow rates were 9.5 m/s, 10 m/s, 10.5 m/s, and 11 m/s. The nozzle's height from the liquid surface was 20 mm, 22.5 mm, 25 mm, 27.5 mm, and 30 mm. Two sets of variables were used for orthogonal experiments, while single parameters were used for parallel experiments, and the results are shown in Figure 12.

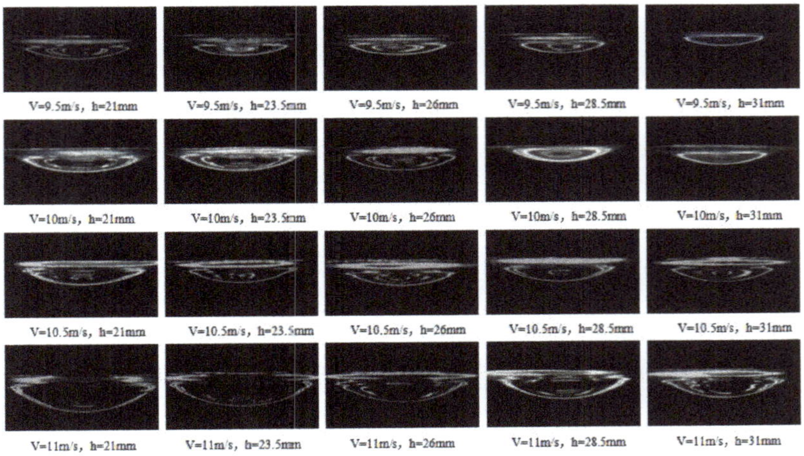

Figure 12. Experimental results of the gas jet forming.

The surface shape of the mirrors was measured using a WILSON MMD-100JS profilometer in Guangzhou, China to obtain the diameter data of the surface formed by the gas jet. The principal curvature of the surface was calculated using the fitting method described in Section 3.4 to determine the principal curvature of the formed surface. The experiments were repeated three times for each set of experimental parameters, and the data obtained are shown in Figure 13.

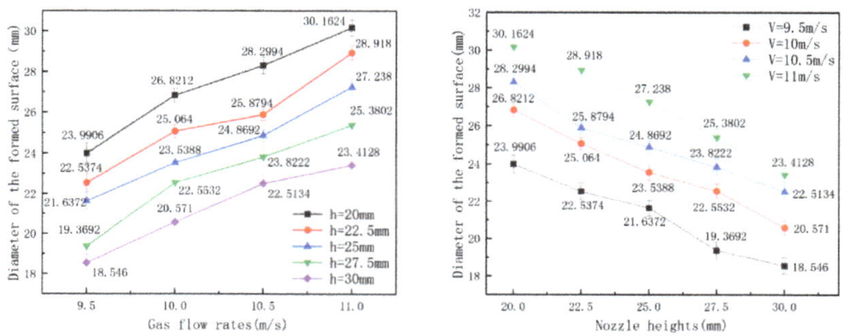

(a) (b)

Figure 13. The diameter of the formed surface with different gas flow rates and nozzle heights in the experiments. (**a**) is the relationship between the gas flow rate and the diameter of the formed surface. (**b**) is the relationship between the nozzle height and the diameter of the formed surface.

The experimental data with principal curvature for the gas jet parameters calculated by fitting the aspheric Equation (19) are shown in Figure 14.

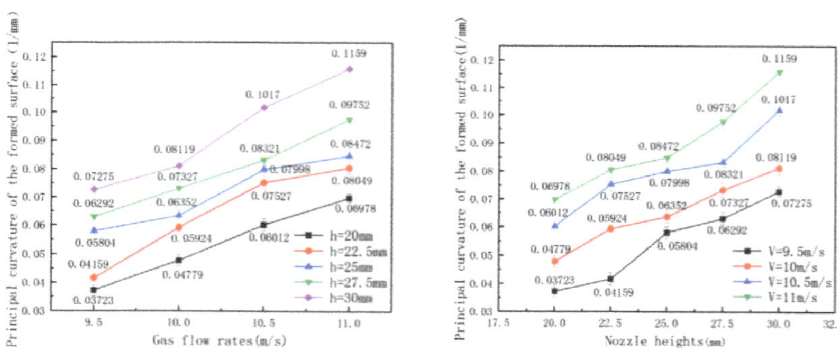

(a) (b)

Figure 14. The principal curvature of the formed surface with different gas flow rates and nozzle heights in the experiments. (**a**) is the relationship between the gas flow rate and the principal curvature of the formed surface. (**b**) is the relationship between the nozzle height and the principal curvature of the formed surface.

6. Discussion

The experimental results show that the diameter of the surface formed by the gas jet increases as the gas flow rate increases and decreases as the nozzle height increases, which is consistent with the conclusions obtained above. The experimental data were compared with the numerical simulation data, and the deviations from the surface shape data obtained are shown in Table 3.

As shown in Table 3, by comparing the numerical simulations with the experimental results, it can be seen that the maximum error was 2.0708 mm and the average error was 1.07 mm for both sets of data.

Table 3. Errors between numerical simulations and experimental data for the diameter of the formed surface.

Parameters	Error	Parameters	Error
$V_0 = 9.5$ m/s, $H = 20$ mm	1.8402	$V_0 = 10.5$ m/s, $H = 20$ mm	1.4042
$V_0 = 9.5$ m/s, $H = 22.5$ mm	1.8428	$V_0 = 10.5$ m/s, $H = 22.5$ mm	0.7813
$V_0 = 9.5$ m/s, $H = 25$ mm	1.5873	$V_0 = 10.5$ m/s, $H = 25$ mm	0.5734
$V_0 = 9.5$ m/s, $H = 27.5$ mm	0.0746	$V_0 = 10.5$ m/s, $H = 27.5$ mm	0.6478
$V_0 = 9.5$ m/s, $H = 30$ mm	0.1444	$V_0 = 10.5$ m/s, $H = 30$ mm	1.6308
$V_0 = 10$ m/s, $H = 20$ mm	2.0708	$V_0 = 11$ m/s, $H = 20$ mm	0.7312
$V_0 = 10$ m/s, $H = 22.5$ mm	0.8694	$V_0 = 11$ m/s, $H = 22.5$ mm	1.3644
$V_0 = 10$ m/s, $H = 25$ mm	1.0398	$V_0 = 11$ m/s, $H = 25$ mm	0.9884
$V_0 = 10$ m/s, $H = 27.5$ mm	1.2586	$V_0 = 11$ m/s, $H = 27.5$ mm	1.1418
$V_0 = 10$ m/s, $H = 30$ mm	0.8166	$V_0 = 11$ m/s, $H = 30$ mm	0.5922

Through the experimental data, it can be seen that the principal curvature of the mirror blank increases with the increase in the nozzle height and increases with the increase in the gas flow rate. The changing law of the principal curvature of the mirror blank caused by the variations of the gas jet parameters is consistent with the numerical simulations. The experimental data were compared with the numerical simulation data, and the deviations from the surface shape data obtained are shown in Table 4.

Table 4. Errors between the numerical simulations and experimental data for the principal curvature of the formed surface.

Parameters	Error	Parameters	Error
$V_0 = 9.5$ m/s, $H = 20$ mm	0.01164	$V_0 = 10.5$ m/s, $H = 20$ mm	0.02202
$V_0 = 9.5$ m/s, $H = 22.5$ mm	0.00086	$V_0 = 10.5$ m/s, $H = 22.5$ mm	0.02211
$V_0 = 9.5$ m/s, $H = 25$ mm	0.00826	$V_0 = 10.5$ m/s, $H = 25$ mm	0.02522
$V_0 = 9.5$ m/s, $H = 27.5$ mm	0.00130	$V_0 = 10.5$ m/s, $H = 27.5$ mm	0.06659
$V_0 = 9.5$ m/s, $H = 30$ mm	0.00561	$V_0 = 10.5$ m/s, $H = 30$ mm	0.07670
$V_0 = 10$ m/s, $H = 20$ mm	0.00824	$V_0 = 11$ m/s, $H = 20$ mm	0.06482
$V_0 = 10$ m/s, $H = 22.5$ mm	0.00323	$V_0 = 11$ m/s, $H = 22.5$ mm	0.07881
$V_0 = 10$ m/s, $H = 25$ mm	0.01480	$V_0 = 11$ m/s, $H = 25$ mm	0.08948
$V_0 = 10$ m/s, $H = 27.5$ mm	0.00882	$V_0 = 11$ m/s, $H = 27.5$ mm	0.09778
$V_0 = 10$ m/s, $H = 30$ mm	0.01875	$V_0 = 11$ m/s, $H = 30$ mm	0.10790

As shown in Table 4, by comparing the numerical simulations with the experimental results, it can be seen that the maximum error was 0.1079 mm^{-1} and the average error was 0.03665 mm^{-1} for both sets of data.

The analysis between the experimental data and the simulation results shows that these two methods have the same trend. However, the numerical simulation has a significant deviation in predicting the surface formed by the gas jet, and the reasons for this deviation are as follows:

Based on the momentum and mass equations, it is clear that the accuracy of the numerical simulation predictions is closely related to the rate properties within the grid. The formula for the rate property of the VOF method is as follows:

$$\mathbf{U}_{\mathrm{VOF}} = \frac{\alpha_{\mathrm{gas}}\rho_{\mathrm{gas}}\mathbf{U}_{\mathrm{gas}} + \alpha_{\mathrm{liq}}\rho_{\mathrm{liq}}\mathbf{U}_{\mathrm{liq}}}{\alpha_{\mathrm{gas}}\rho_{\mathrm{gas}} + \alpha_{\mathrm{liq}}\rho_{\mathrm{liq}}}, \tag{20}$$

where α_{gas} is the gas phase volume fraction, ρ_{gas} is the gas phase physical density, $\mathbf{U}_{\mathrm{gas}}$ is the gas phase lattice rate, $\boldsymbol{\alpha}_{\mathrm{liq}}$ is the liquid phase volume fraction, ρ_{liq} is the liquid phase physical density, and $\mathbf{U}_{\mathrm{liq}}$ is the liquid phase lattice rate.

$$\mathbf{U}_{error} = \mathbf{U}_{\mathrm{VOF}} - \mathbf{U}_{\mathrm{liq}} = \frac{\alpha_{\mathrm{gas}}\rho_{\mathrm{gas}}\mathbf{U}_{\mathrm{gas}} + \alpha_{\mathrm{liq}}\rho_{\mathrm{liq}}\mathbf{U}_{\mathrm{liq}}}{\alpha_{\mathrm{gas}}\rho_{\mathrm{gas}} + \alpha_{\mathrm{liq}}\rho_{\mathrm{liq}}} - \mathbf{U}_{\mathrm{liq}}, \tag{21}$$

During the formation process, the difference in rate between the gas phase compared to the liquid phase is large. Differentiating the VOF method rate attributes $\mathbf{U}_{\mathrm{VOF}}$ and $\mathbf{U}_{\mathrm{liq}}$ yields the following final formula:

$$\mathbf{U}_{error} = \frac{\alpha_{\mathrm{gas}}\rho_{\mathrm{liq}}\left(\mathbf{U}_{\mathrm{gas}} - \mathbf{U}_{\mathrm{liq}}\right)}{\alpha_{\mathrm{gas}}\rho_{\mathrm{gas}} + \alpha_{\mathrm{liq}}\rho_{\mathrm{liq}}}, \tag{22}$$

In the VOF method, the gas and liquid phases must fill a certain grid. This means that the gas and liquid phases are in the control body, and the volume fraction sums to 1. The formula is as follows:

$$\alpha_{\mathrm{gas}} + \alpha_{\mathrm{liq}} = 1, \tag{23}$$

The mirror blanks were formed at a constant temperature of 20 °C, at which point the density of air is 1.225 $\mathrm{kg/m^3}$ and the density of the liquid phase is 977 $\mathrm{kg/m^3}$. Equation (24) was obtained by substituting the density data into Equation (22).

$$\mathbf{U}_{error} = \frac{977\alpha_{\mathrm{gas}}\left(\mathbf{U}_{\mathrm{gas}} - \mathbf{U}_{\mathrm{liq}}\right)}{\alpha_{\mathrm{gas}}1.225 + \alpha_{\mathrm{liq}}977}, \tag{24}$$

In the VOF method, the accuracy of the subtype prediction in this test case is related to the difference between the mixed-phase rate property $\mathbf{U}_{\mathrm{VOF}}$ and the associated $\mathbf{U}_{\mathrm{liq}}$. After deduction, the factors influencing $\mathbf{U}_{error}$ include α_{gas}, α_{liq}, and the difference in rate $\mathbf{U}_{\mathrm{gas}} - \mathbf{U}_{\mathrm{liq}}$ between the gas and liquid phases. Assuming that within a specific grid at a certain α_{gas} and α_{liq} the time step is constant in this paper, the gas–liquid two-phase rate is the only factor that affects them. Since the liquid phase is a highly viscous fluid and the gas phase is air, the difference between the two phases is too significant, causing iterative errors.

7. Prediction Modelling of Surface Shape Parameters

The sizeable computational error at the interface of the gas–liquid phase in the numerical simulation is hard to correct, making it difficult to accurately predict the surface shape of the gas-jet-formed surface. Based on this issue, we accurately predicted the formed surface shape by establishing a dimensionless prediction model of the gas jet parameters and the surface shape parameters. The model was built by matching equations based on the variation trend of the experimental data and fitting the equation coefficients using the BFGS method. The prediction model for the formed surface diameter is shown in Equation (25).

$$d_{re} = \beta_1 + \beta_2/V_0{}^2 + \beta_3 \mathrm{Ln}h, \tag{25}$$

where V_0 is the gas flow rate at the gas nozzle (m/s), h is the nozzle height (mm), and d_{re} is the surface diameter of the mirror blank (mm).

The BFGS method was used to fit the equation coefficients, and the results are shown in Table 5.

Table 5. Coefficients of the prediction model for the formed surface diameter.

β_1	β_2	β_3
Density/kg·m^{-3}	1.225	977

The gas jet parameters were set to V_0 = 9.8 m/s, V_0 = 10.3 m/s, h = 22 mm, and h = 24.5 mm for the orthogonal experiments, and the following conclusions can be obtained by comparing the experimental data with the calculation results of the surface diameter prediction model established in this paper: The maximum deviation of the experimental results was 0.8554 mm, and the average deviation was 0.3192 mm. The prediction accuracy of the model was much better than that of the numerical simulation.

The same method was used to build a predictive model for the principal curvature, which takes the form shown in Equation (26).

$$k = \alpha_1 + \alpha_2 v + \alpha_3 {V_0}^2 + \alpha_4 h^{\alpha_5}, \tag{26}$$

where V_0 is the initial rate at the gas nozzle (m/s), h is the nozzle height (m), and k is the principal curvature of the mirror blank (mm^{-1}). The BFGS method was used to fit the equation coefficients, and the results are shown in Table 6.

Table 6. Coefficients of the prediction model for the principal curvature of the formed surface diameter.

α_1	α_2	α_3	α_4	α_5
−0.49287	0.07184	−0.00231	8.28075	1.28306

The gas jet parameters were set to V_0 = 9.8 m/s, V_0 = 10.3 m/s, h = 22 mm, and h = 24.5 mm for the orthogonal experiments, and the following conclusions can be obtained by comparing the experimental data with the calculation results of the principal curvature prediction model established in this paper: The maximum deviation of the experimental results was 0.00735 mm^{-1}, and the average deviation was 0.00269 mm^{-1}. The prediction accuracy of the model was much better than that of the numerical simulation.

8. Conclusions

In this paper, the pressure distribution at the potential core region of the gas jet was analysed theoretically and combined with numerical simulations to draw the following conclusions: The formed surface diameter increases as the flow rate increases and decreases as the nozzle height increases. The principal curvature increases with increasing flow rate and nozzle height. According to the test results, when the nozzle's height from the liquid surface is 20–30 mm, the jet velocity is 9.5–11 m/s, the minimum value of the surface diameter formed by the gas jet is 18.5460 mm, and the maximum value is 30.1624 mm. The minimum value of the principal curvature formed by the gas jet is 0.03723 mm^{-1}, and the maximum value is 0.1159 mm^{-1}. The prediction of the surface parameters of the gas jet forming by the numerical simulation method was found to have large deviations in the experiments. The experimental results were compared with the simulation predictions, and it was found that the average prediction deviation of the diameter was 1.07 mm. The average prediction deviation for the principal curvature was 0.03665 mm^{-1}, which is challenging to correct in simulation. Therefore, we developed a dimensionless prediction model of the surface curvature and surface diameter of the mirror blanks by considering the jet parameters using experimental data. The model's average prediction error of the diameter of the formed surface was 0.3192 mm, and the average prediction error of the principal curvature of the formed surface was 0.00269 mm^{-1}—better than the numerical simulation.

Author Contributions: W.F.: conceptualisation, methodology, software, investigation, writing—original draft. M.W.: validation, formal analysis, visualisation, software. X.Z.: validation, formal analysis, visualisation, investigation. All authors have read and agreed to the published version of the manuscript.

Funding: This research was funded by the "Basic and Applied Basic Research Fund of Guangdong Province, grant number 2021A1515110927", the "Jilin Province Scientific and Technological Development Program, grant number 20200201006JC", "The Open Project Program of Key Laboratory for Cross-Scale Micro and Nano Manufacturing, Ministry of Education, Changchun University of Science and Technology, grant number CMNM-KF202108", and the "Scientific Research Project of Education Department of Guangdong Province, grant number 2022KCXTD029".

Institutional Review Board Statement: Not applicable.

Informed Consent Statement: Informed consent was obtained from all subjects involved in the study.

Data Availability Statement: Not applicable.

Conflicts of Interest: The authors declare no conflict of interest.

References

1. Martin, H.M.; Allen, R.G. Production of 8.4 m segments for the giant magellan telescope. *SPIE Astron. Telesc. Instrum.* **2012**, *8450*, 801–815.
2. Ma, Z.; Peng, L.; Wang, J. Ultra-smooth polishing of high-precision optical surface. *Optik* **2013**, *124*, 6586–6589. [CrossRef]
3. Jedamzik, R.; Werner, T.; Westerhoff, T. Production of the 4.26 m ZERODUR mirror blank for the Advanced Technology Solar telescope (ATST). *SPIE Astron. Telesc. Instrum.* **2014**, *9151*, 1012–1020. [CrossRef]
4. Westerhoff, T.; Schäfer, M.; Thomas, A.; Weissenburger, M.; Werner, T.; Werz, A. Manufacturing of the ZERODUR 1.5-m primary mirror for the solar telescope GREGOR as preparation of light weighting of blanks up to 4-m diameter. *SPIE Astron. Telesc. Instrum.* **2010**, *7739*, 206–214.
5. Hatefi, S.; Abou-El-Hossein, K. Review of single-point diamond turning process in terms of ultra-precision optical surface roughness. *Int. J. Adv. Manuf. Technol.* **2019**, *106*, 2167–2187. [CrossRef]
6. Liu, Y.; Xing, Y.; Yang, C.; Li, C.; Xue, C. Simulation of heat transfer in the progress of precision glass molding with a finite element method for chalcogenide glass. *Appl. Opt.* **2019**, *58*, 7311–7318. [CrossRef] [PubMed]
7. Fu, W.; Zhang, X. Numerical simulation assisted curve compensation in spin casting of optical mirror blank. *Optik* **2020**, *207*, 164406. [CrossRef]
8. Fu, W.; Zhang, X. Experimental and numerical simulation study on gas jet forming of optical mirror blank. *Optik* **2020**, *207*, 164451. [CrossRef]
9. Banks, R.B.; Bhavamai, A. Experimental study of the impingement of a liquid jet on the surface of a heavier liquid. *J. Fluid Mech.* **1965**, *23*, 229–240. [CrossRef]
10. Molloy, N.A. Impinging jet flow in a two-phase system: The basic flow pattern. *J. Iron Steel Inst.* **1970**, *208*, 943–950.
11. Meidani, A.R.N.; Isac, M.; Richardson, A.; Cameron, A.; Guthrie, R.I.L. Modelling Shrouded Supersonic Jets in Metallurgical Reactor Vessels. *Iron Steel Inst. Jpn.* **2004**, *44*, 1639–1645. [CrossRef]
12. Ek, M.; Sichen, D. Study of Penetration Depth and Droplet Behavior in the Case of a Gas Jet Impinging on the Surface of Molten Metal using Liquid Ga–In–Sn. *Steel Res. Int.* **2012**, *83*, 678–685. [CrossRef]
13. Kalifa, R.B.; Mahjoub Saïd, N.; Bournot, H.; Le Palec, G. Numerical Study of a Gas Jet Impinging on a Liquid Surface. In *International Conference Design and Modeling of Mechanical Systems*; Springer: Cham, Switzerland, 2017; pp. 661–670.
14. Kofman, N.; Mergui, S.; Ruyer-Quil, C. Characteristics of solitary waves on a falling liquid film sheared by a turbulent counter-current gas flow. *Int. J. Multiph. Flow* **2017**, *95*, 22–34. [CrossRef]
15. Rosler, R.S.; Stewart, G.H. Impingement of gas jets on liquid surfaces. *J. Fluid Mech.* **1968**, *31*, 163–174. [CrossRef]
16. Vouros, A.P.; Panidis, T. Turbulent properties of a low Reynolds number, axisymmetric, pipe jet. *Exp. Therm. Fluid Sci.* **2013**, *44*, 42–50. [CrossRef]
17. Mi, J.; Nobes, D.S.; Nathan, G.J. Influence of jet exit conditions on the passive scalarfield Of an axisymmetric free jet. *J. Fluid Mech.* **2001**, *432*, 91–125. [CrossRef]
18. Banks, R.B.; Chandrasekhara, D.V. Experimental investigation of the penetration of a high-rate gas jet through a liquid surface. *J. Fluid Mech.* **1963**, *15*, 13–34. [CrossRef]
19. Hirt, C.W.; Nichols, B.D.; Romero, N.C. SOLA: A Numerical Solution Algorithm for Transient Fluid Flows. In *Energy Research and Development Administration*; Energy Research and Devempment Administration: Washington, DC, USA, 1975; pp. 1–19.
20. Nichols, B.D.; Hirt, C.W.; Hotchkiss, R.S. *SOLA-VOF: A Solution Algorithm for Transient Fluid Flow with Multiple Free Boundaries*; Los Alamos National Lab: Los Alamos, NM, USA, 1980. [CrossRef]
21. Ishii, M.; Mishima, K. Two-fluid model and hydrodynamic constitutive relations. *Nucl. Eng. Des.* **1984**, *82*, 107–126. [CrossRef]
22. Brackbill, J.; Kothe, D.; Zemach, C. A continuum method for modeling surface tension. *J. Comput. Phys.* **1992**, *100*, 335–354. [CrossRef]
23. Jeon, S.S.; Kim, S.J.; Park, G.C. CFD Simulation of Condensing Vapor Bubble using VOF Model. *World Acad. Sci. Eng. Technol.* **2009**, *60*, 209–215.
24. Evans, M.R.; Oliver, D.; Shekhar, S.; Harvey, F. Fast and Exact Network Trajectory Similarity Computation: A Case-study on Bicycle Corridor Planning, ACM SIGKDD. In Proceedings of the 2nd ACM SIGKDD International Workshop on Urban Computing, Chicago, IL, USA, 11 August 2013; Association for Computing Machinery: New York, NY, USA, 2013; pp. 1–8.
25. Fréchet, M.M. Sur Quelques Points du Calcul Fonctionnel. *Rend. Del Circ. Mat. Di Palermo* **1906**, *22*, 1–72. [CrossRef]

26. Chen, L.; Özsu, M.T.; Oria, V. Robust and Fast Similarity Search for Moving Object Trajectories. In Proceedings of the 2005 ACM SIGMOD International Conference on Management of Data, Baltimcre, MD, USA, 14–16 June 2005; Association for Computing Machinery: New York, NY, USA, 2005; pp. 491–502.
27. Schulz, G. Imaging Performance of Aspherices in comparison with Spherical Surfaces. *Appl. Opt.* **1987**, *26*, 5118–5142. [CrossRef] [PubMed]

applied sciences

Article

Digital Simulation and Analysis of Assembly-Deviation Prediction Based on Measurement Data

Ninglu Zhang, Yingna Wu, Rui Yang and Guangping Xie *

Center for Adaptive System Engineering, School of Creativity and Art, ShanghaiTech University, Shanghai 201210, China; zhangnl@shanghaitech.edu.cn (N.Z.); wuyn@shanghaitech.edu.cn (Y.W.); yangrui@shanghaitech.edu.cn (R.Y.)

* Correspondence: xiegp@shanghaitech.edu.cn

Abstract: For various power equipment items such as aircraft engines and gas turbines with numerous components demanding requirements on processing accuracy and high structural complexity, processing errors tend to cause such problems as assembly rework, forced assembly and prolonged research and development cycle. In this paper, the core machine of micro gas turbine is taken as the research object to construct a simulation model of assembly-deviation prediction based on the design tolerance and actual measurement data. Then, an analysis is conducted on the assemblability of the design model and the key factors causing the deviation of the assembly. After conducting deviation calculations and simulation analysis, it was determined that the current mounting position is deemed to be suboptimal. In light of this finding, an optimized solution is proposed, which involves advancing the mounting phase by 0.47 mm. This adjustment effectively resolves the interference issue resulting from the design error. Moreover, the geometry, shape and three-dimensional contour of the key components are measured with high accuracy and precision to identify the key characteristic parameters affecting the outcome of the assembly. With an assembly-deviation-prediction and -analysis model established on the basis of actual measurement data, the results of the assembly-deviation analysis are compared with the outcome of the assembly, showing high consistency. The assembly-deviation-prediction method proposed in this paper on the basis of design tolerance and actual measurement data is applicable to the manufacture of aviation and combustion engines.

Keywords: digital assembly; tolerance; deviation analysis

Citation: Zhang, N.; Wu, Y.; Yang, R.; Xie, G. Digital Simulation and Analysis of Assembly-Deviation Prediction Based on Measurement Data. *Appl. Sci.* **2023**, *13*, 12193. https://doi.org/10.3390/app132212193

Academic Editors: Ivan Galvão, Rui Manuel Leal and Gustavo H. S. F. L. Carvalho

Received: 10 May 2023
Revised: 27 June 2023
Accepted: 3 July 2023
Published: 10 November 2023

1. Introduction

Aircraft engines, gas turbines and other power equipment components are characterized by a large number of components, complex shape and structure, demanding requirements on processing accuracy, a complex assembly process and so on, as commonly encountered in the manufacturing process. At the design stage, the lack of scientific evaluation system for component tolerance allocation or unreasonable tolerance allocation affects the accuracy of the assembly and the assemblability of the components. At the processing stage, the commonly used traditional one-dimensional and two-dimensional methods of measurement tend to ignore the impact of three-dimensional error. However, the assembly requires a perfect match between three-dimensional parts, which means that even the correct one-dimensional or two-dimensional size does not ensure that the parts can be successfully assembled. At the assembly stage, aircraft engines, gas turbines and other sophisticated equipment have demanding requirements on the accuracy of assembly for working in high-temperature and high-pressure environments [1]. When the processed component exceeds the allowable range of tolerance, it can result in rework that incurs unnecessary time and financial costs. Usually, assembly time accounts for about 20% to 50% of the total production time [2]. Forced assembly occurs even more frequently, which has a significant impact on the quality, performance and lifespan of gas turbines and other important equipment. In some cases, it even causes damage to all the equipment [3].

Assembly is realized in a three-dimensional space, where there are many 3D chains present in non-parallel planes, as well as the geometric deflation effects of angular tolerances and transfer coefficients. A 3D chain represents a kind of complex nonlinear dimensional chain. By establishing a mathematical model for the deviation of the 3D dimensional chain with space vectors, Zhang [4] analyzed the impact caused by the deviation of each constituent ring on the deviation of the closed ring according to the sensitivity matrix in it. Meng et al. [5] established the vector ring model of the 3D chain through the concept of vector and then applied it to build the "dichotomous" 3D model of deviation simulation.

Analytical verification of assemblability and simulation verification of assembly accuracy and performance are the two primary components of digital assembly. Developed countries place profound value on digital assembly. For instance, Boeing has employed model-based design (MBD) in the development of the 787 airliner to achieve rapid development under three-dimensional digital design and simulation technology, leading to a significant reduction in assembly process-planning time [6]. Companies such as GE, Pratt & Whitney and Rollo have optimized the original aero-engine rotor assembly simulation technology with augmented reality technology, resulting in a 50% reduction in the error rate of workers assembling the harness and a 25% reduction in completion time [7]. The digital assembly of aircraft engines and gas turbines has garnered immense attention from numerous scholars. Shi [8] has made strides in this area, by successfully establishing a diagnostic mathematical model of core machine performance, and subsequently analyzing the errors of significant measurement parameters associated with the core machine. Liu et al.'s [9] contributions to this field include the proposal of a dual-objective optimization principle, utilizing the Monte Carlo method to optimize rotor noncentricity and unevenness. Shan [10] has made remarkable headway by introducing the method of Pearson distribution family, which has been instrumental in the modeling and analysis of assembly deviation within typical rotor parts found in aero-engines. Similarly, Wang et al. [11] have made notable contributions, obtaining integrated runout results based on component-measurement outcomes, and determining optimal assembly angles via optimization calculations. Gao [12] proposed a core engine rotor assembly accuracy-prediction model utilizing the concept of elastic contact of the assembly surface. By adjusting and optimizing the installation phase of the rotor based on the model predictions, the assembly accuracy was significantly improved. Chen et al. [13] developed an assembly-prediction method by leveraging high-precision big data collection and "assembly deformation compensation and correction" technology. This method allowed for the prediction of the optimal assembly angle and position of each stage of the rotor disk, thereby improving assembly accuracy and reducing the potential for errors.

Hillyard [14] first proposed the use of computers to define the geometric shape, dimensions and tolerances of parts. With the progress in computer-aided tolerance analysis, there have been more and more studies conducted on digital assembly via 3D tolerance-analysis software. Zhang [15] conducted a study on dimensional deviations in engine compartments using 3DCS software and performed a root cause analysis of the resulting assembly deviations. Meng et al. [16] adopted the VSA software to investigate the cause of significant concentricity of pivot point for a type of core machine test part. Zheng et al. [17] used 3DCS software to analyze and optimize the tolerance of spacing in the landing gear intersection for the nacelle of the turboprop engine, which solved the problem with intersection spacing of the landing gear. The aforementioned studies present practical instances and methodologies for the analysis and optimization of assembly utilizing deviation-analysis software. However, these studies solely focus on the analysis and optimization of assembly deviations in design models, without further integrating real-world components with the design models. In addition, there is no guidance provided for actual assembly processes through the utilization of digitized assembly based on real-world components.

The assembly is realized in a 3D space, where a small amount of discrete 1D and 2D measurement data cannot reflect the 3D shape, the distribution of high and low points, and the exact location of the component. Therefore, for the assembly-deviation analysis of the machined parts, it is necessary to combine 1D, 2D and 3D measurement methods and to integrate the obtained data to better indicate the size, shape, deformation and contour error of the machined parts.

To perform 1D and 2D measurement of parts with high precision, various high-precision measurement equipment items such as coordinate-measuring machines, laser trackers and articulated arm coordinate-measuring machines can be used. Moreover, for the assembly of ultra-large equipment, high-precision digital photogrammetry can be used as well. In terms of 3D measurement, the high-precision 3D laser scanning equipment based on line laser, structured light and phase shift method is advantageous. Currently, many researchers are investigating and applying 3D measurement methods. For instance, Yu [18] employed binocular grating 3D measurement technology to measure an impeller, achieving a low-cost and high-accuracy 3D graphic visualization of the impeller. Tan et al. [19] utilized structural light to measure the surface differences of a body-in-white in 3D and addressed the issues of poor accuracy and low efficiency of manual measurement by fitting the reference line using the RANSAC algorithm. Lao et al. [20] utilized various 3D measurement instruments and proposed a 3D measurement process model-construction method for aircraft assembly.

In this paper, micro gas turbine is taken as the research object to establish a digital assembly-deviation-prediction and -analysis model based on design tolerance and actual measurement data. This not only improves the rationality of design and the reliability of assembly but also accelerates the process of equipment development and reduces production costs, which demonstrates its applicability in practice.

2. Digital Simulation Prediction and Analysis Method of Assembly Deviation

Figure 1 illustrates the digital prediction and analysis method developed in this paper for detecting the occurrence of assembly deviation, which involves the digital simulation of assembly deviation based on design tolerance, the in-machine 3D digital measurement of components and the digital simulation of assembly deviation based on actual measurement data.

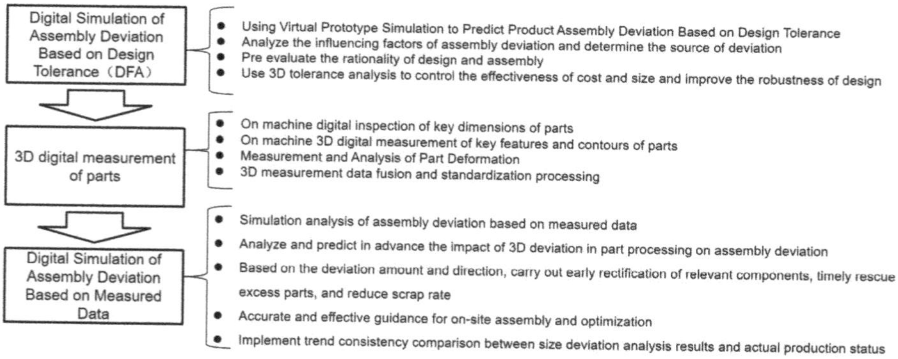

Figure 1. Digital assembly process.

Digital simulation of assembly deviation based on design tolerance.

Based on design tolerance, the digital simulation model of assembly deviation is established in 3DCS, and the Monte Carlo method is used to simulate the assembly at least 2000 times to obtain the prediction results of the assembly deviation. The simulation results are used to analyze the key influencing factors for assembly deviation, determine the cause of assembly deviation and evaluate the rationality of the design before actual production. If the design is reasonable, the parts of the machine can be produced, and if it is not reasonable, the design needs to be changed until the design meets the requirements.

By adopting advanced 3D tolerance analysis, cost and dimensional effectiveness can be controlled, improving design robustness.

Three-dimensional digital measurement of components.

By identifying the key influencing factors for assembly deviation and the causes of error through design tolerance analysis, the key dimensions, form errors and 3D contours of key features related to the assembly of the actual machined components are subjected to highly accurate measurement, 3D scanning. In this paper, the MetraSCAN 750TM | Elite handheld laser 3D scanning system is applied in combination with Hexagon's GLOBAL CMM for the measurement of key components of the core machine. High-precision dimensional measurement of parts is mainly carried out via CMM, and part contour scanning and part deformation measurement is carried out via a handheld scanner. The scanned data were processed in a standardized manner and extracted using Polyworks. Then, the part measurement data and the design model were fused in 3DCS to create a fusion model of the part.

The digital simulation of assembly deviation based on actual measurement data.

Based on the actual measurement data of the assembled parts, the assembly-deviation-simulation model was established in 3DCS to predict and analyze the impact of machining errors on assembly deviation. If the simulation results of assembly deviation meet the design requirements, the parts can be assembled into a machine, and if they do not meet the design requirements, the parts need to be repaired. By taking into account the extent and direction of assembly deviation as measured, the relevant parts with excessive deviation can be reworked in advance, which reduces the scrap rate and improves the efficiency of assembly. Moreover, the accuracy of on-site assembly can be ensured and the visual guidance and optimization of on-site assembly can be achieved.

3. Digital Simulation Prediction and Analysis Experiment of Assembly Deviation

3.1. Simulation Modeling and Analysis of Assembly Deviation Based on Design Tolerance

In the prediction and analysis model of assembly-deviation simulation based on design tolerance, the Monte Carlo method is used to carry out assembly simulation several times by transferring the changes of each element in the chain in the form of random numbers for the key characteristic tolerances of each component affecting the outcome of the assembly. At each time of simulation, the assembly-deviation data are collected for statistical calculation and analysis of the pattern and characteristics of distribution of assembly deviation, as required to evaluate the assembly qualification rate and to analyze tolerance allocation. Thus, the key factors affecting the accuracy of assembly and the main cause of error are determined.

In this paper, the assembly of the core engine used for micro gas turbine is studied through a verification experiment for the digital simulation and analysis of assembly deviation. The core engine consists of various key components such as compressor, combustion chamber and turbine, as shown in Figure 2. The turbine of the micro-combustion engine studied in this paper is a centripetal radial turbine, the assembly of which is required to ensure a sufficient clearance between the rotor blade tip and the guide of the core engine, so as to prevent the rotating parts from collision with the stationary parts during operation. All relevant dimensional features in the dimensional chain where the clearance is located are important. In this paper, the 3DCS tolerance simulation software is used to model and simulate the assembly deviation of the design model, with the simulation results referenced to assess the reliability and rationality of the design.

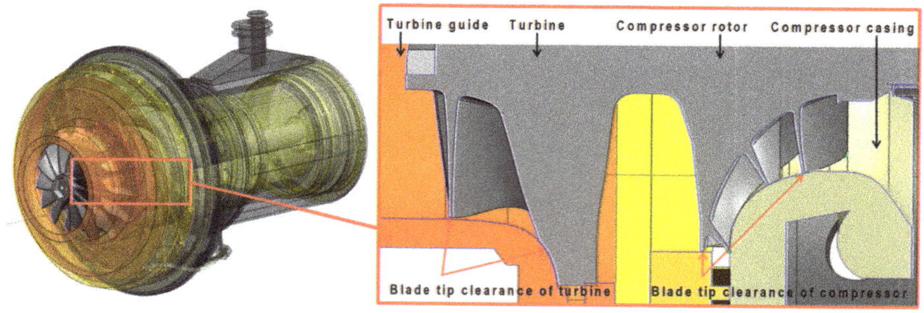

Figure 2. Schematic diagram of rotor assembly.

The process of modeling the simulation of assembly deviation based on design tolerance is as follows:

1. Import the assembly model of the part into 3DCS.
2. Establish part tolerances, construct feature meshes and select the appropriate feature tolerance scores according to the machining of the component, such as normal distribution, skewed distribution, Rayleigh distribution, etc.
3. Define the assembly scheme. The non-rotating parts of the motor and core machine are stacked assembly, and the rotor is a rotating part mounted on it by two bearings. The non-rotating parts are subjected mainly to stacked assembly, through flange connection, stop positioning, etc. The suitable assembly methods, such as feature assembly and two-point assembly, are adopted to construct a lightweight model for the assembly of the core machine.
4. Establish the measurement target according to the assembly requirements. As shown in Figure 3, it is difficult to perform mesh fitting for the complex curved parts such as the core machine rotor, and the measurement must be converted into the measurement of distance between discrete points. In this paper, the rotor blade tip surface is replaced by 16 blade tip key position points, and the cassette surface is changed into a mapping point for the blade tip point, with 16 measurement targets defined. In this way, the simulation error is significantly reduced.
5. Perform simulation and analysis optimization. Two thousand Monte Carlo assemblies are performed on the design model of the part to obtain its digital assembly results. Figure 4 shows the simulation results of an assembly gap-measurement target, with the probability of assembly interference reaching 28.75%. It is indicated that the reliability and rationality of the design are not satisfactory, which means the assembly must be optimized. Therefore, the assembly interference is solved in this paper by modifying and optimizing the position in which the rotor is mounted. The simulation reveals that the assembly no longer causes interference and the gap is uniform when the mounting position of the rotor is moved forward by 0.47 mm, as shown in Figure 4b.

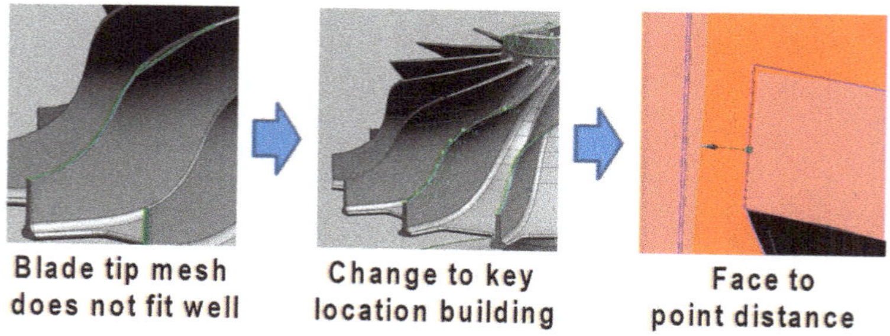

Figure 3. Discrete point distance-measurement model.

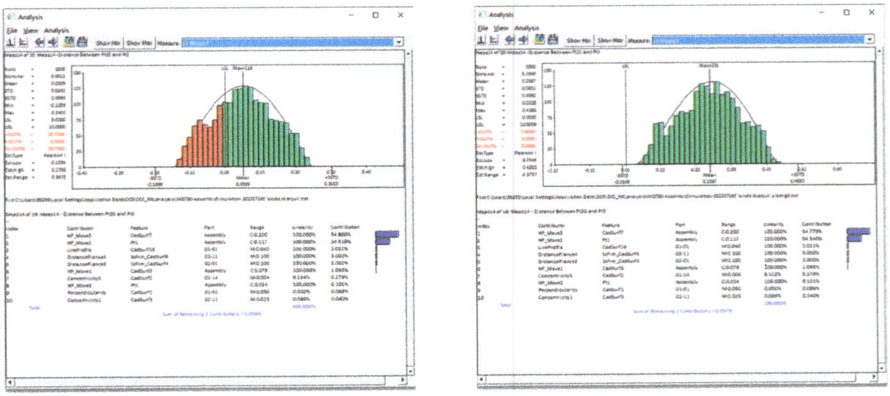

(a) (b)

Figure 4. Calculation results of measurement point 14 of the micro-turbin: (**a**) Before assembly optimization; (**b**) After assembly optimization.

3.2. Precision Measurement and Data Processing of Key Components

3.2.1. Three-Dimensional Precision Measurement of Key Components of the Core Machine

In this paper, the MetraSCAN 750TM | Elite handheld laser 3D scanning system is applied in combination with Hexagon's GLOBAL CMM for high-precision 1D, 2D and 3D measurement of key components of the core machine. Figure 5 shows how CMM and 3D scanning work.

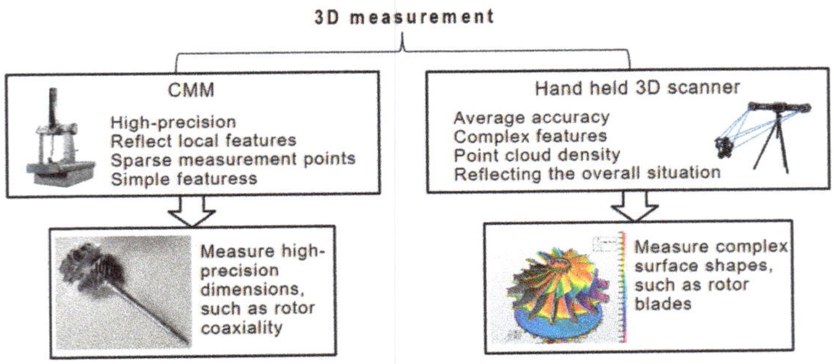

Figure 5. The division of labor and cooperation between CMM and 3D scanning.

The assembly of the micro-combustion engine is required to ensure a sufficient clearance between the rotor blade tip and the guide of the core engine, so all relevant feature dimensions along the dimensional chain where the clearance is located need to be measured.

The dense point cloud data obtained via laser 3D scanning reflect the 3D shape, the distribution of high and low points along with the exact location of the parts, etc., which makes it suitable for the scanning of complex and curved parts. There are many micro gas turbines with complex curved parts suitable for data extraction via 3D scanning, such as the compressor impeller and turbine impeller used in the combustion engine rotor. Herein, it is verified by repeated scanning of the standard ball with the handheld laser 3D scanning system that the instrument deviation is within 0.025 mm, which meets the need to measure the parts with a less demanding accuracy requirement than 0.05 mm. For the parts with smooth surfaces, the quality of the point cloud obtained via the laser scanner can be compromised. To solve this problem, such methods as powder coating of the part surface can be used to obtain high-quality point clouds [21]. To measure, the part is fixed to the fixture and then scanned with a 3D scanner to obtain all the point cloud data of the part. To avoid errors in the scanning results, each part needs to be scanned three times. For those large and complex parts, it is usually necessary to scan the part from multiple perspectives and then combine the point cloud data obtained from multiple angles to determine the complete 3D shape of the part.

CMM is characterized by sparse measurement points and high measurement accuracy, with the highest accuracy reaching up to 0.001 mm, which makes it suitable for the highly accurate measurement of dimensional and tolerances. There are many key components in the core machine requiring high processing accuracy; for example, the roundness, coaxiality and straightness of the combustion engine rotor must be within 0.004 mm. In this case, the measurement can be performed via the CMM with higher measurement accuracy. To measure, the workpiece is fixed on the CMM platform and then the probe is used to measure high-precision part assembly features with accuracy requirements within 0.04 mm.

3.2.2. 3D Scanning Point Cloud Data Processing

Figure 6 shows the process of 3D point cloud data processing for those key parts obtained via the 3D laser scanner, which mainly involves the collocation of multi-angle 3D scanned point cloud data, point cloud data denoising and the segmentation and feature extraction of point cloud data according to the needs for assembly features. In the processing of point cloud data, it is also a common practice to repair the holes of the point cloud. For the 3D point cloud needed to examine the actual processing quality of the parts, point cloud repair can lead to data distortion and affect the outcome of assembly simulation.

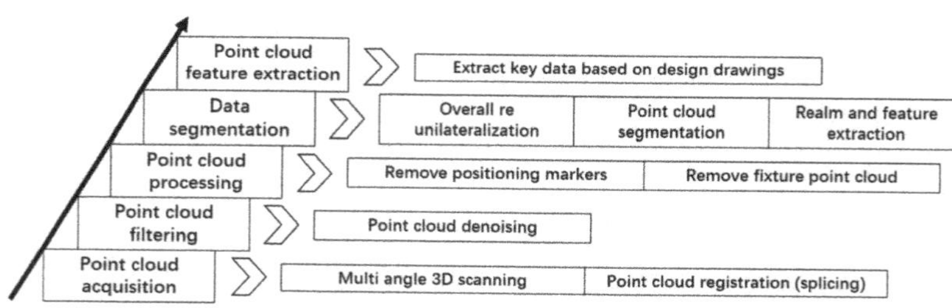

Figure 6. Point cloud data-processing flowchart.

To extract the key part feature parameters, conversion of the measured coordinate system into the design coordinate system of the part is usually required. For the parts produced with design models, the best fit can be aligned with the CAD model to create a complete picture of the error between the machined part and the CAD design model. The data of dimensions, form, deformation and contour of key features are collected by extracting the actual point cloud measurement data at the corresponding location of the CAD model, creating features based on the drawings of the part, extracting the key features.

To extract the parameters of key features from 3D scanned point cloud data, there are different methods available, such as probing, fitting, CAD model-based extraction, etc. The probing method employed in this study bears resemblance to a virtual CMM technique. Within the software, the scanned point cloud is treated as a physical entity, and a specific feature to be measured is selected. Subsequently, a virtual probe is employed to detect the point cloud, thereby determining the corresponding feature and its associated measurement value. On the other hand, the fitting method utilized in this research involves fitting the scanned data to standard features. To accomplish this, the point cloud is subjected to a least squares fitting approach using Polyworks software. Finally, the CAD model extraction method is an automated technique that leverages a designated region surrounding the corresponding feature in the CAD model as the actual measurement data area. This enables the direct extraction of the measurement data. The accuracy achieved with different methods varies. Take the mounting hole of the core machine studied in this paper as an example. In design, the diameter of the mounting hole is 170.43 mm, and the actual CMM value of the part is 170.593 mm. With the CMM data as the standard value, the data are extracted by using different methods. Then, the data obtained via different extraction methods are compared, the results of which are shown in Table 1.

Table 1. Measurement results of mounting hole (mm).

Serial Number	Probing Method	Fitting Method	CAD Method
1	170.584	170.602	170.637
2	170.586	170.595	170.637
3	170.568	170.603	170.637
4	170.590	170.610	170.637
5	170.538	170.583	170.637
Average value	170.573	170.599	170.637
Difference	−0.012	0.006	0.044

According to the findings presented in Table 1, the outcomes derived from the probing method exhibit the most minimal disparity when compared to the CMM values. This can be attributed to the enhanced capability of the probing method to comprehensively capture the intricate details of the features by utilizing a dense set of points. Thus, the presence of stray points is effectively eliminated during this process, thereby leading to an improved accuracy in the extraction of feature parameters. On the other hand, the probing method relies on sparsely distributed points, which fails to provide a holistic representation of the entire feature, resulting in substantial deviations in the measured values. Similarly, the CAD method, despite its automated extraction capability, is susceptible to erroneous selection of point clouds and extraction ranges, thereby leading to significant deviations in the measured values.

3.3. Modeling and Analysis of Assembly-Deviation Prediction Based on Actual Measurement Data

The assembly of the micro gas turbine core studied in this paper is representative of stacked assembly. Ideally, the axes of different parts are on the same reference axis after assembly, and the combined end face is a plane perpendicular to the cylindrical axis. However, the machined parts are flawed, and there may be various errors in form, thus resulting in a three-dimensional distortion caused to stacked assembly, as shown in Figure 7. To establish the assembly-deviation-prediction and -analysis model based on actual measurement data, it is necessary to extract the dimensional data, form tolerance, axis data and end face data of the machined parts. Then, these measurement data should be inputted into the assembly-deviation-prediction and -analysis model as established in Section 3.1 based on the design tolerance. Meanwhile, the replacement of the design value of the feature with the actual measured value is necessary. This is particularly crucial for the feature that extracts the actual data through point cloud scanning. In this case, the design tolerance must be substituted with the measurement uncertainty of the handheld scanning instrument, which amounts to 0.025 mm. In the validation experiments conducted in this paper, the axial data of the part are extracted through least squares fitting of the cylindrical surface point cloud data related to the assembly, while the three-dimensional point cloud data of the end face of the part are extracted through least squares fitting of the surface and three high points constituting the plane, respectively. Depending on the method of end face data extraction, two simulation models of assembly deviation are established on the basis of measurement data and then compared with the results of assembly to explore the extraction method that can produce the most comparable outcome to the actual results.

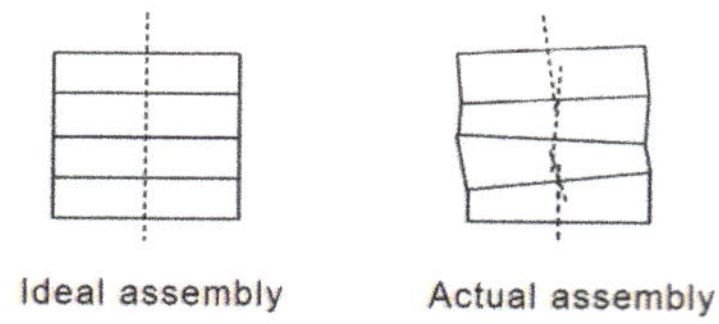

Figure 7. Ideal assembly and actual assembly.

3.3.1. Simulation Modeling of Assembly Deviation Based on Least Squares Fitting Surface

For the obtained 3D point cloud data, all the point clouds that form a certain plane in the scanned data are selected and a fitting plane is created via the least squares method. In Figure 8, Ls LCS indicates the least squares fitting plane. To obtain the assembly-deviation model based on the least squares fitting plane, the geometric center and normal vector data of the least squares fitting plane are introduced into the design model.

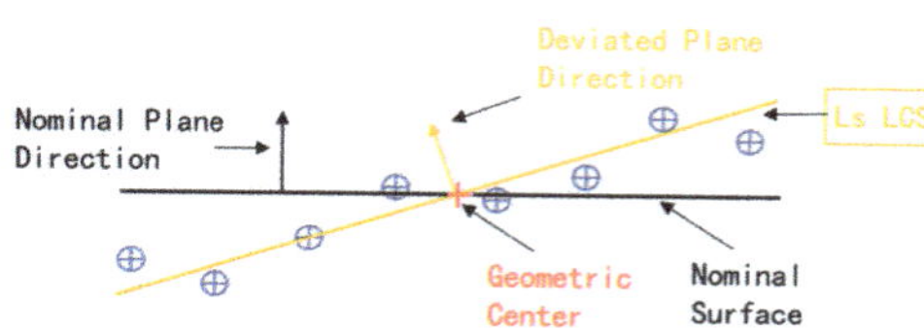

Figure 8. Least squares fitting surface of plane features.

3.3.2. Simulation Modeling of Assembly Deviation Based on Three High Point Composition Planes

Although the least squares fitted plane can reflect the characteristics of the bonding surface in general, the machined assembly plane is usually irregular in shape and characterized by "double high points" or "multiple high points". Apart from that, there are significant deviations arising between the least squares fitted plane and the actual plane shape, which affects the accuracy of predicting the tilt of the stacked assembly axis. Figure 9 shows the deviation color map of the processing plane point cloud data and the ideal plane. Compared with the ideal plane, point clouds with warmer colors are more prominent, while point clouds with colder colors are more concave. As shown in Figure 10, according to the deviation color map of the point cloud data, three representative high points that do not share the same line are searched for and a plane is established to reflect the tilt of the assembly. With the three high point fitting plane as the assembly plane, an assembly-deviation-simulation model is established on the basis of the plane formed by the three high points.

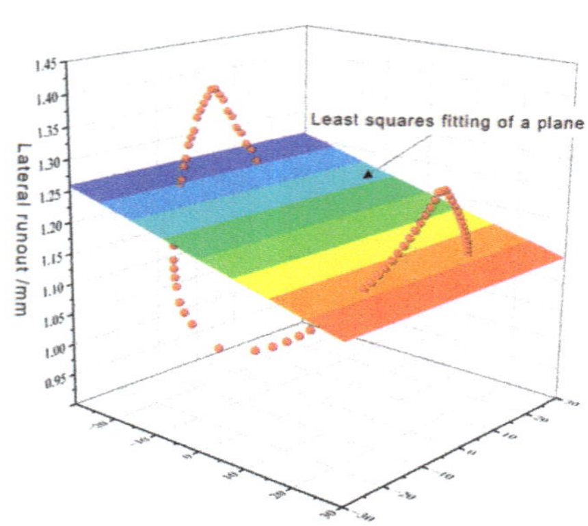

Figure 9. Least squares fitting plane of "double high point" plane.

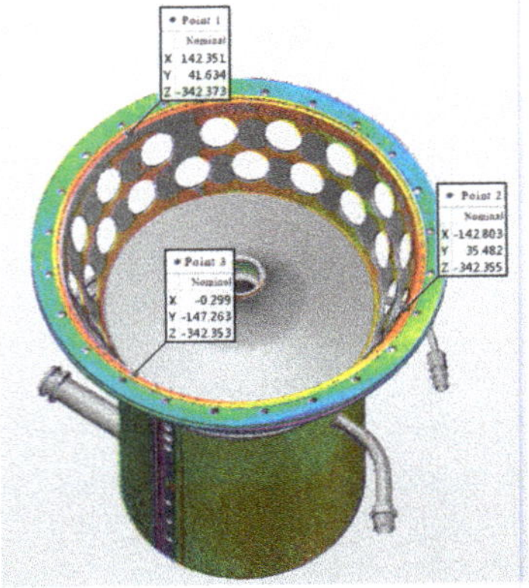

Figure 10. Extraction of representative three points on the mounting surface of motor housing.

3.3.3. Comparison of the Analysis Results of Two Assembly-Deviation-Simulation Models

For the assembly of the core part of the micro gas turbine, the total radial clearance of the outermost clearance of the rotor turbine tip and turbine guide is 1.2 mm, as determined by the measurement without interference and collision; the radial runout of the rotor blade tip is 0.183 mm. When the rotor is pushed backwards by 0.47 mm, the pressurizer rotor collides with the pressurizer magazine. Two-thousand Monte Carlo assemblies are carried out on the fusion model of part design and measurement to obtain the digital assembly results. The simulation results obtained through the two assembly-simulation models are compared with the measurement results, the results of which are shown in Table 2.

Table 2. Comparison of simulation results of different assembly methods.

	A	B	C
Actual assembly	1.2 mm	0.183 mm	100.0%
Assembly Based on Least Squares Fitting Surface	1.2 mm	0.194 mm	67.3%
Assembly Based on Three High Points Forming a Plane	1.2 mm	0.202 mm	99.5%

A—Radial total clearance between the turbine blade tip and the outermost clearance of the turbine guide. B—Radial runout of rotor blade tip. C—Collision probability of 0.47 mm pushed back.

As can be seen from Table 2, the simulation results of the assembly deviation based on the three high points of the composition plane are comparable to the measurement results of the assembly, which is conducive to the prediction of assembly deviation.

4. Discussion

The assembly-deviation-simulation model, based on three high points constituting a plane, considers the highest point of the point cloud constitution as the mounting surface, which is a closer approximation to the actual state of the axial length superposition than the least squares fitting model. As such, the prediction for the collision is more consistent with the actual state. However, this method employs fewer points, which results in a steeper slope of the fitted surface than that of the least squares fitted surface. Consequently, there may be a greater slope in the radial direction of the assembly model, leading to significant deviations in the predicted radial runout values.

When comparing the results with previous studies, it is important to note that the old studies provided practical examples and methods for analyzing and optimizing assembly utilizing deviation-analysis software [15–17]. Based on these examples and methods, a simulation model was successfully developed to predict assembly deviations using design tolerances and measured data, focusing on a micro gas turbine as the research object. However, the previous studies only focused on the analysis and optimization of assembly deviations in design models, without further integrating the measured data of the components with the design models. In this study, highly accurate measurements of the geometric shape, form and three-dimensional contour of key components were performed to determine the critical feature parameters that affect assembly results. An assembly-deviation-prediction and -analysis model was established based on the actual measurement data, and a comparison was made between the results of the assembly-deviation analysis and the assembly results to identify the best digital assembly-deviation-prediction model based on actual measurements. This research analyzes and predicts the assembly of actual components through a digital assembly based on measured data, providing a basis and guidance for the actual assembly process.

Although there are important discoveries revealed by these studies, there are also limitations. Further research can be conducted in the future on the following:

1. One avenue for further research involves exploring the digital assembly of thin-walled and easily deformable parts. While this study analyzed the micro gas turbine parts as rigid components, it is acknowledged that certain components within the turbine are thin-walled and deformable. As such, the established digital assembly model has inherent limitations. Expanding research efforts to encompass the assembly of these thin-walled and deformable parts could yield a more accurate simulation and analysis model of assembly deviation. Furthermore, such research has the potential to offer valuable insights for the assembly of advanced power plants, including aviation and combustion engines.
2. Additional investigation into the mathematical model of assembly is required. Several scholars have proposed novel mathematical models for digital assembly and deviation analysis. Further research on mathematical models can facilitate the integration of theory and practice, leading to the establishment of a digital assembly-deviation-analysis model that more accurately reflects the real-world scenario.
3. An in-depth simulation study is necessary to investigate the thermal deformation of core machine parts. As the core machine operates under high-temperature and high-pressure conditions, accurate predictions of its thermal assembly are crucial. To obtain the deformation of parts at high temperatures, factors such as thermal expansion coefficient, elastic modulus, density and deformation curve corresponding to the high-temperature deformation of the material can be assigned in CAE software based on the working temperature distribution of the core machine. This study can provide a better understanding of the thermal behavior of core machine parts and inform the development of more reliable and accurate thermal assembly models.

5. Conclusions

In this paper, the assembly of micro gas turbine core engines is taken as the research object to establish an assembly-deviation-analysis model based on design tolerance. Moreover, the key factors for the occurrence of assembly deviation and the causes of error are analyzed through Monte Carlo operations, which solves the assembly interference and collision caused by unreasonable design through optimization. This adjustment effectively resolves the interference issue resulting from the design error, indicating that digital simulation of assembly deviation based on design tolerances will help industry and researchers reduce design errors in practice. The assembly-deviation-analysis model is established on the basis of measurement data and different feature parameter-extraction methods. Then, the results of simulation analysis of assembly deviation based on the least squares fitting plane and the three high point fitting plane are compared with the actual results of assembly deviation. It is found that the simulation results of assembly deviation based on the three high point fitting plane are basically consistent with the actual outcome of the assembly, and the three high point fitting will help industry and researchers predict assembly deviations.

The method proposed in this paper to predict and analyze the assembly deviation based on the design tolerance and measurement data is applicable in practice.

Notwithstanding its contributions, this study is not without certain limitations, such as analyzing the micro gas turbine parts as rigid components, not further analyzing the deformation of thin-walled parts, not further investigating the mathematical model of assembly and not further analyzing the assembly deformation in the thermal state, which can be further investigated in the future.

Author Contributions: Conceptualization, R.Y.; methodology, G.X.; software, N.Z.; validation, G.X. and N.Z.; data curation, N.Z.; writing—original draft preparation, N.Z.; writing—review and editing, G.X.; supervision, Y.W.; project administration, R.Y. and Y.W.; All authors have read and agreed to the published version of the manuscript.

Funding: This research received no external funding.

Institutional Review Board Statement: Not applicable.

Informed Consent Statement: Not applicable.

Data Availability Statement: No new data were created or analyzed in this study. Data sharing is not applicable to this article.

Conflicts of Interest: The authors declare that they have no known competing financial interest or personal relationship that could have appeared to influence the work reported in this paper.

References

1. Lyu, Y.H.; Wu, F.H.; Wei, X.P.; Shi, X.Y. Summarization of Concentricity Measurement Research in Engine Assembly Progress. In Proceedings of the 15th Annual Conference of China Association for Science and Technology: Symposium on Aero-Engine Design, Manufacturing and Application Technology, Guiyang, China, 25 May 2013.
2. Yu, H.T.; Mao, B.; Luo, Z.W. Digital Preassembly Detection System For Aircraft Assembly Process. *Aeronaut. Manuf. Technol.* **2015**, *4*, 55–60. [CrossRef]
3. Liu, J.; Wu, F.Y.; Wang, J. Optimization Technique of Aeroengine Rotor Assembly. *Aeroengine* **2014**, *40*, 75–78. [CrossRef]
4. Zhang, L. Deviation Analysis of 3-D Dimension Chain. *Mech. Eng.* **2015**, *7*, 105–107. [CrossRef]
5. Meng, Q.F.; Zhang, L.X.; Yin, L.Z.; Hou, Z.X.; Zou, F. Three Dimension Tolerance Analysis and Optimization Based on 3DCS. *J. Syst. Simul.* **2018**, *30*, 1730–1738. [CrossRef]
6. Liu, Z.J.; Liu, Z.W.; Wu, X.F.; Hu, S.J.; Wei, X.G. Research on key technologies of aero-engine digital assembly simulation. *Aerodynamics* **2021**, *2*, 74–76.
7. Liu, X.B. Research on MBD-Based Aircraft Assembly Process Auxiliary System. Master's Thesis, Shenyang Aerospace University, Shenyang, China, 2013.
8. Shi, Y. Preliminary study on measurement error of important performance parameters of a core computer. *J. Shenyang Aerosp. Univ.* **2011**, *28*, 12–16+26.
9. Liu, J.; Wu, F.Y.; Wang, J. Aeroengine rotor assembly optimization technology. *Aircr. Engine* **2014**, *40*, 75–78.
10. Dan, F.P.; Li, Z.M.; Zhu, B. Modeling and analysis of assembly deviation of typical aeroengine rotor parts. *Manuf. Autom.* **2015**, *37*, 100–103.
11. Wang, B.; Li, S.G. Application of assembly optimization technology in the assembly of a certain type of aeroengine high pressure compressor rotor. *Aviat. Maint. Eng.* **2016**, *6*, 51–53.
12. Gao, Y.C. Research on Prediction of Assembly Accuracy and Phase Optimization of Aeroengine Rotor. Master's Thesis, Dalian University of Technology, Dalian, China, 2020.
13. Chen, K.; Tang, X.L.; Ye, F.; Li, J.; Deng, S.P.; Wang, P.; Ning, Y. Research on simulation and prediction of aeroengine rotor assembly process. *Fan Technol.* **2021**, *63*, 72–78.
14. Hillyard, R.C. Dimensions and Tolerances in Shape Design. Ph.D. Thesis, University of Cambridge, Cambridge, UK, 1978.
15. Zhang, L. Analysis and Research on Dimension Deviation of Engine Room Compartment. Master's Thesis, Hefei University of Technology, Hefei, China, 24 April 2015.
16. Meng, X.H.; Cheng, X.Y.; Shan, F.P. Geometric Tolerance Analysis of a Certain Type Core Engine. *Manuf. Autom.* **2018**, *40*, 74–78+85. [CrossRef]
17. Zheng, W.; Chen, J.X.; Guo, C.P.; Yu, C.X.; Xu, J.L.; Chen, X.T. Deviation Analysis and Tolerance Optimization of Space between Landing Gear Assembly Joint in Turboprop Engine Nacelle. *Aeronaut. Manuf. Technol.* **2022**, *65*, 87–93. [CrossRef]
18. Yu, L.W. Study on the Structure Light Measurement of Centrifugal Pump Impeller. Master's Thesis, Wuhan Institute of Technology, Wuhan, China, 15 May 2017.
19. Tan, Y.; Yan, H.Y.; Yang, Y.Z. Three-dimensional Measurement Method for Surface Difference of White Body Based on Structured Light. *Tool Eng.* **2022**, *56*, 87–90. [CrossRef]

20. Lao, C.J.; Zhao, Z.Y.; Sun, S.H.; Liu, Q.D. Constraction Technology of 3D Measurement Process Model for Aircraft Assembly. *China Meas. Test* **2020**, *46*, 98–101.
21. Xu, Y.M.; Tian, L. Precision Testing and Analysis of HandySCAN 700. *J. Geomat.* **2021**, *46*, 55–58. [CrossRef]

MDPI AG
Grosspeteranlage 5
4052 Basel
Switzerland
Tel.: +41 61 683 77 34

Applied Sciences Editorial Office
E-mail: applsci@mdpi.com
www.mdpi.com/journal/applsci

www.ingramcontent.com/pod-product-compliance
Lightning Source LLC
Chambersburg PA
CBHW061556210926
51850CB00009B/2119
* 9 7 8 3 7 2 5 8 2 0 7 6 4 *